# The Backyard Sheep

# The BACKYARD SHEEP

An Introductory Guide to Keeping
Productive Pet Sheep

Sue Weaver

Storey Publishing

*The mission of Storey Publishing is to serve our customers by
publishing practical information that encourages
personal independence in harmony with the environment.*

Edited by Deborah Burns and Lisa H. Hiley
Art direction and book design by Cynthia N. McFarland

Cover photography © Jason Houston, except author's photograph © John Weaver
Interior photography and ephemera courtesy of the author: 1, 2, 4, 9, 34, 65, 66, 113,
    114, 134, 153, 185, and 203; and © Barry Bland/Minden Pictures, 156
Illustration and color section photography credits appear on page 204

Indexed by Susan Olason

Thanks to Debbie Dane and Hallie Walker for their assistance with the cover photography

**Storey Publishing**
210 MASS MoCA Way
North Adams, MA 01247
*www.storey.com*

Printed in the United States by Versa Press
10  9  8  7  6  5  4  3  2  1

LIBRARY OF CONGRESS CATALOGING-IN-PUBLICATION DATA on file

This book is dedicated to Baasha and Baamadeus,

two great sheep who have crossed the Rainbow Bridge

— you are sadly missed, my woolly friends —

and to my husband, John,

who helps keep things running smoothly.

# Contents

# Preface

Baa-ram-ewe, baa-ram-ewe.
To your breed, your fleece, your clan be true.
Sheep be true. Baa-ram-ewe.

— The sheep's password from the movie *Babe* (1995)

I came to sheep late in life, in 2002, after moving from east-central Minnesota to the Arkansas Ozarks. Friends in Minnesota kept sheep; I enjoyed working with them, but until our move I'd never owned any. Then I met Anita Messenger of Liberty Mountain Ranch in Bismarck, Arkansas. She knew of two old sheep who needed a home and I had a home that needed sheep. So it was that Dodger, a Hampshire, and Angel, a Wiltshire Horn cross, came to us.

I quickly became besotted with sheep. I wanted more, and I wanted to raise lambs. I began shopping for a breed and narrowed my choices to Karakuls, Scottish Blackface, and Cheviots. An Internet search led me to Linda Coats of Coats High Ridge Farm in Columbia, Missouri, who was dispersing her flock of Miniature Cheviots.

Correspondence ensued and with it came pictures. I fell in love with her middle-aged foundation ewe; a month later, Brighton Ridge Farms #59, a black, beautiful matron I renamed Baasha, joined Dodger and Angel, along with a black ram lamb I called Wolf Moon Finvarra (Abram to his friends). More little sheep followed: two of Baasha's daughters and a granddaughter — and they all had lambs, beautiful lambs that stole my heart.

One day while filling the horse tanks I chanced to notice a scrap of wool snagged on the fence. At that moment, in a rush of emotion, I felt a deep connection with thousands of years of shepherds to whom wool and sheep meant life. Sheep fed and clothed ancient people. Sheep hides gave them leather and parchment. Sheep carried packs and pulled wagons. Their manure made fallow ground bloom. Women fashioned soap from sheep tallow, and sheep tallow candles lighted the night. Gut strings gave voice to musical instruments; sheep horns became shofars (wind instruments used to communicate battle and announce religious ceremonies).

Sheep no longer hold the central role in life they held before our modern world evolved; nonetheless, easy-to-care-for, affordable sheep still have much to offer. Consider the variety of wools available to fuel today's handspinning and fiber art renaissance; the ultrarich, high-fat milk to craft fine cheese; the luscious lamb for meat eaters; and the reason for and means with which to train and trial herding dogs. Best of all, they are peerless small-farm pets ready and willing to entertain their owners with their beauty, personality, and joie de vivre.

Can you tell how much I love sheep?

# Part One

# Learning about Sheep

a, Welsh Sheep; b, South-Down Sheep; c, Dorset Sheep; d, Black-faced Cheviot Sheep; e, Norfolk Sheep; f, Ryland Sheep.

# CHAPTER 1

# Sheep throughout History

*Sheepe doth both with his fleece apparrell us, and with his milke and wholesome flesh nourish us.*

— Barnaby Googe, *The Whole Art and Trade of Husbandry* (1614)

The Navajo name for sheep, *dibah*, means "that by which we live," but the Navajo are not unique in their dependence on sheep. Throughout recorded history sheep meant life to mankind. Even prior to domestication, wild sheep provided meat to sustain hunter societies and pelts to keep them warm. Sheep (along with their cousins, goats) were simple to domesticate, becoming part of the family soon after dogs were domesticated.

## Why Sheep?

Sheep neatly fit the criteria for domestication as espoused by anthropologist Sir Francis Galton, cousin of Charles Darwin, in *The First Steps towards the Domestication of Animals* (a paper published in *The Reader*, Volume 2, December, 1863).

They are hardy. Their young survive if removed from their parents and raised by man. Wild neonatal lambs could be cornered and snatched and raised on human milk. That's probably how domestication began.

Sheep that are raised by humans like people. Sheep society is based on a dominance hierarchy; sheep follow and defer to humans who raise them.

They prefer to live in close proximity to their own kind. You don't need a lot of room to keep sheep.

Sheep breed freely in captivity. They are small and easy to take care of, reasonably easygoing, and versatile in their feeding habits. They flock together and can be tended by one or two individuals.

## Sheepish Origins

While the exact ancestors of domestic sheep remain unknown, the Asian Mouflon (Ovis aries orientalis) is certainly involved. Five subspecies of Mouflon, a species that historically abounded throughout the areas where sheep were domesticated, share the same chromosome count (54) and DNA material as today's domestic sheep.

Two to three thousand years before the birth of Christ, Neolithic farmers settled the isolated four-island archipelago we now call St. Kilda, located 41 miles (66 km) off the western coast of Scotland. They brought with them semi-wild sheep much like the Asian Mouflons from which domestic sheep descend.

They salted the least habitable of the four islands with their primitive sheep to better utilize the sparse grazing available in St. Kilda's fierce, storm-swept climate. In this hostile environment, the little sheep thrived. (When Viking raiders visited in the seventh and eighth centuries CE, they named the island So-øy, meaning "Sheep Island" after the nimble, diminutive sheep dwelling there. )

Soay are (the plural of Soay is Soay) a living remnant of these early domesticated sheep. These lithe, lean, fine-boned, elfin animals are active, sure-footed, and nimble. They have coarse hair coats in lieu of obvious wool; short, skinny tails; and horns. They grow an incredibly soft undercoat of wool as winter approaches and shed it again in the spring, exactly in the manner of their wild ancestors.

### DID EWE KNOW?

#### Where the Wild Sheep Are

Today, true wild sheep populations occur in western North America (Bighorn Sheep), northern Canada and Alaska (Dall's Sheep), Siberia (Snow Sheep), Afghanistan and Pakistan (Urial), eastern Asia (Argali), western Asia (Asian Mouflon), and Corsica and Sardinia (European Mouflon).

Most researchers, however, believe European Mouflon are the feral descendants of domesticated sheep taken to Europe by Neolithic Middle Eastern farmers around 7000 to 6000 BCE.

Asian Mouflon

Argali sheep

Wild sheep occurred in shades of brown; most species have dark upper bodies and lighter bellies, a pattern that renders them as inconspicuous as possible when standing in full sunlight. Rams (males) have huge curved or twisted horns; in most species ewes (females) have horns, too. Their tails are shorter than those of domestic sheep.

The major difference, however, is their body coat: wild sheep grow coarse outer hair, not wool. In fact, wild sheep grow a precursor to wool; it's the soft, downy winter undercoat that keeps them warm. Under domestication, early humans selected for more undercoat and less hair until today's woolly sheep evolved.

## Sheep, Sheep, Everywhere

Archaeozoologists routinely unearth bones from domesticated sheep at ruins as diverse as 11,000-year-old Zawi Chemi Shanidar in the foothills of the Zagros Mountains in northern Iraq; Suberde, a tiny Turkish village nestled in a mountain basin in southwestern Anatolia, occupied between 7400 and 7000 BCE; and Pre-Pottery Neolithic B Jericho, dating to 6800 BCE. We know domestication occurred in Asia Minor between 8,000 and 10,000 years ago, but not precisely when or where.

Within a few thousand years, domestic sheep had become shorter legged and smaller than their wild kin. Their tails grew longer while their horns grew shorter and their skulls shrank. New colors emerged, among them white sheep (a most desirable color because white wool readily accepts natural dyes). And, perhaps most significant in their symbiotic relationship to humans, they stopped shedding their increasingly longer and woollier

coats. Longer fleeces could be shorn and thus harvested all at once, and they provided longer fiber that was more easily spun into yarn.

At first, women worked fleece by hand, rolling tufts of fleece along their thighs, adding more fiber as needed to create a length of yarn. Later, the drop spindle was developed. Stone and clay whorls (weights) from ancient drop spindles have been recovered at Middle Eastern sites dating to 5000 BCE. All processed wool was spun with some variation of the handheld spindle until the spinning wheel was invented in the thirteenth century CE. Spindles are still used throughout the world today.

Woolly, nonshedding sheep evolved early on. Archaeologists working a dig near Sarab, Iran (a site noted today for its geometric-patterned Persian wool rugs), unearthed a figurine of a woolly sheep carbon-dated to 4000 BCE. The Greek poet Homer wrote of the fine white wool produced in Thessaly, Arcadia, and Ithaca, while Pliny the Elder, writing in the first century CE, heaps praise on wool produced in the Greek coastal city of Tarentum.

But sheep meant more to ancient man than meat, pelts, and wool. A Mesopotamian clay tablet incised in 2000 BCE documents a farmer's production of sheep's milk, butter, and cheese; in addition, sheep hides were used for crafting parchment. The oldest surviving sheepskin parchment document is the Leviticus Scroll of the Dead Sea Scrolls, written between the second century BCE and the first century CE. The *Codex Vaticanus* and *Codex Siniaticus*, two surviving copies of the entire *New Testament* dating from the fourth century CE, are written on sheepskin parchment.

## Sheep in Great Britain

Authorities believe that sheep came to Britain around 4000 to 3000 BCE as Neolithic farmers from Europe traveled across the English Channel. At the time of the Roman conquest in 43 CE, the weaving of woolen fabric was already well established in Britain.

According to the Doomsday Survey ordered by William I in 1086 to record England's assets and liabilities, there were more sheep in England than all other livestock species combined. Henry II fined British woolmen (workers who packed raw wool) for forming a craft guild in 1180 without gaining his prior approval. That guild evolved into the Worshipful Company of Woolmen, incorporated in 1522 and still in existence today.

William the Conqueror altered the course of Britain's history in 1271 when he invited skilled Flemish weavers to immigrate and establish a textile industry in England. Then he placed an export tax on all wool shipped to Europe. This tax ("the Great Custom") was the first of many enacted to control Britain's immense wool trade. By 1300, there were an estimated 15 million sheep in the British Isles.

Wool was medieval England's major export. Medieval abbeys produced wool on a staggering scale and grew rich through its sale. When Henry VIII issued the Dissolution of the Monasteries in 1537, however, the baton passed to wool merchants who organized the delivery of raw fleece to workers and collected the finished pieces, which they delivered to wool marketing centers such as London and Southampton.

By 1660, two-thirds of England's foreign commerce involved textiles and wool. As wool merchants in the Cotswolds and East Anglia grew rich, they built lavish estates and donated funds that helped build some of the finest churches in the land; edifices that are still known as the "wool churches."

### DID EWE KNOW?

### A Mountain of Bones

In 2002, while working on a survey of ancient sites on Wiltshire's Salisbury Plain, British archaeologist David McOmish discovered a 2,600-year-old ritual feasting area near present-day East Chisenbury. Established on the site of an even older abandoned settlement, it encompasses a hill comprising the bones of at least 500,000 sheep.

## BRITISH SHEEP TERMINOLOGY

While most sheep terms, such as *ewe, ram,* and *lamb,* are the same in all English-speaking parts of the world, some are specific to given regions, for example Britain, where you may encounter these terms:

**CADE, PODDY.** A hand-reared orphan or bottle lamb

**CLAGS, CLARTS, DAGS.** Manure hanging in fleece, particularly on the hindquarters

**COUPLES.** Ewes with their lambs

**DRAW LAMBS.** Assist with lambing

**FLYBLOWN.** To suffer from fly strike (maggots in the flesh)

**FOLD.** A pen in which sheep are kept overnight to keep them safe from predators

**GIMMER, THEAVE.** A ewe lamb from weaning to first shearing

**HOG, HOGG, HOGGET, HOGGERAL.** A sheep of either sex between weaning and first shearing

**MULE.** A hybrid produced by crossing sheep of two different breeds

**SHEARLING.** A sheep of either sex between its first and second shearings

**SHORNIE.** A recently shorn sheep

**SUCKER.** A suckling lamb

**TUP.** A ram

**TUPPING.** The mating of sheep

**WEDDER.** A wether, a castrated male sheep

**WIGGING.** Trimming the fleece from around a sheep's eyes to prevent wool blindness

**YOW.** Regional colloquialism for ewe

And here are some surnames that come from the wool trade:

**BOWKER.** Washed finished cloth

**CARD, CARDER, TOZER, TOWZER, KEMP, KEMPER, KEMPSTER.** Pertaining to combing the wool

**CHAPMAN, MERCER.** A wool merchant

**CROPPER, FOSTER, SHEARMAN, SHERMAN.** One who sheared sheep

**DRAPER.** One who prepared woolen cloth and sold it to the tailors

**DYER, DEXTER, LISTER.** One who dyed the wool

**FULLER, WALKER, WELCKER, TUCK, TUCKER, TUCKERMAN.** One who "fulled" (felted) wet, soapy cloth by trampling on it

**PACKER, PACKMAN, LANE, LANEY, LANIER.** One who transported fleece and finished fabric

**SHEPHERD, SHEPLY, SHEPPEY, TUPPER.** One who tended sheep

**STAPLER, STAPLES.** One who purchased raw wool

# Counting Sheep

Shepherds who lived in Britain prior to the Industrial Revolution had a unique way of counting sheep called *Yan Tan Tethera*. Shepherds counted their flocks several times a day: first thing in the morning and last thing at night for sure, but also when performing routine tasks like shearing and trimming hooves. The systems they used are based on an ancient Celtic language, probably Cumbric, but different ones evolved in various parts of Britain. All were based on the number 20. To count more than 20 sheep, the shepherd would count to 20, then make a mark on the ground and start again. Others cut grooves in their crooks or staffs and moved their thumbs to a new mark with each lot of 20 sheep.

Yan Tan Tethera is still used in some parts of Britain; this one is used in Swaledale in northern Yorkshire.

| Number | Word | Number | Word | Number | Word |
| --- | --- | --- | --- | --- | --- |
| 1 | Yan | 8 | Akker | 15 | Bumfit |
| 2 | Tan | 9 | Counter | 16 | Yanabum |
| 3 | Tethera | 10 | Dick | 17 | Tanabum |
| 4 | Mether | 11 | Yanadick | 18 | Tetherabum |
| 5 | Pip | 12 | Tanadick | 19 | Metherabum |
| 6 | Azer | 13 | Tetheradick | 20 | Jiggit |
| 7 | Sezar | 14 | Metheradick | | |

**LEARN FROM THE PAST**

The sheep is frequently employed in the dairy regions of this country [the United States], at the tread-mill or horizontal wheel, to pump the water, churn the milk, or perform other light domestic work.

— R. L. Allen, *Domestic Animals: History and Description of the Horse, Mule, Cattle, Sheep, Swine, Poultry, and Farm Dogs* (1848)

## Sheep in the Americas

Sheep first arrived in Jamestown Colony in 1608 and were eaten during the winter of 1609. Fifteen years after settling Plymouth Colony, the Pilgrims purchased sheep from Dutch dealers on Manhattan Island (though they probably had sheep before then). By 1643 there were 1,000 sheep in the Massachusetts Bay Colony alone. Britain soon after that forbade further importations, but Colonial Americans continued to smuggle in thousands of sheep.

Britain also forbade the formation of an American woolen textile industry. In 1698 the penalty for being involved in America's underground wool trade was having a hand lopped off. Nevertheless, Americans proudly wore garments made of American wool. Presidents George Washington and Thomas Jefferson, both of whom raised sheep at their estates, wore American-made wool suits at their inaugurations. The first woolen factory in the United States was established in 1788 in Hartford, Connecticut.

Spanish explorers brought sheep to the Americas, too. Queen Isabella of Spain financed the voyages of Christopher Columbus and Hernando Cortez with money derived from the Spanish woolen industry. Columbus brought ultrahardy, coarse-wool churro meat sheep to Cuba and Santo Domingo on his second voyage in 1493; in 1519, part of Cortez's walking food supply was composed of the descendants of Columbus's sheep. Spanish conquerors took churros to South America, too, where they replaced the "sheepe of Peru" (llamas) as primary livestock.

## The Rise of the Merino

The vast Spanish wool industry was based on short-stapled, ultrafine-wool Merino sheep, believed to have come to Spain during the twelfth century along with the Moors. By 1526 there were three and one-half million Merino sheep in Spain. They were forbidden from being exported, on pain of death, until 1786, when King Ferdinand VI of Spain sent several hundred fine Merino sheep to his cousin, King Louis XVI of France. Louis installed these valuable woolies at his estate at Rambouillet, where the fine-wool Rambouillet breed was born. Once the floodgates opened, affluent shepherds around the world imported Merinos, which

Spanish Merino sheep

became the finest of the fine-wool breeds, as they still are today.

Ferdinand shipped sheep from his famous Escoriale Merino Stud to the Dutch House of Orange, but they couldn't adjust to the high rainfall in the Netherlands; so in 1789, two rams and six ewes were sent to Colonel Jacob Gordon in Capetown, establishing Merinos in South Africa, where today there are more than 14 million of them.

Australia's first sheep, a fat-tailed meat breed, arrived in 1788 with England's First Fleet. Thirteen Merinos followed in 1797. By 1830, there were nearly two million sheep in Australia, and in 1838 alone, the annual Merino wool clip topped 4.4 million pounds (2 million kilos). It's said that "Australia is built on the sheep's back" because her early wealth came from the sale of Australian-grown wool.

Naturally Merinos eventually came to America and particularly to Vermont, thanks to the efforts of William Jarvis, the American consul at Lisbon in Portugal. In 1809, Spain granted him permission to export two hundred rams from the Escorial royal flock to Boston. He returned home in 1810 and moved to his 2,000-acre farm in Weathersfield, Vermont, with four hundred Merinos. Over the course of his lifetime he imported 3,630 Spanish Merinos, effectively establishing the breed on our shores, though they were most popular in New England and the Midwest. Western farmers used Merino and Rambouillet stock to create American breeds like the Columbia.

## Sheep in the Rest of the World

Sheep entered Africa via Sinai and have been present in Egypt since between 6000 and 5000 BCE. Numerous Egyptian deities were associated with rams, even from the predynastic era; rams were revered for their virility and warlike attributes. Sheep migrated the length of Africa and are depicted among the thousands of stunningly beautiful San Bushmen rock paintings in the Drakensberg Mountains of South Africa.

In 1654, when the Dutch East India Company sent Jan van Riebeeck to South Africa to establish a trading post to supply company ships with fresh victuals as they rounded the Cape, he found that the native Hottentots already had fat-tailed sheep. He traded tobacco,

---

**LEARN FROM THE PAST**

As to the very first importation from Spain to the United States, there seems to be a difference of opinion. One says in 1793 Mr. William Foster of Boston, Mass., imported two ewes and a ram smuggled through the port of Cadiz. These Mr. Foster gave to Mr. Andrew Craigue of Cambridge, who, not knowing their value, killed and ate them. Another says the first importation was by William Porter of Boston in 1798. These he gave to Mr. Andrew Craigue of Cambridge, who, not knowing their value, killed and ate them. One thing seems pretty certain. The first importations of Spanish Merino sheep into the United States disappeared as mutton, and were of no account from a breeding standpoint, but beyond a doubt the mutton was good.

— Dr. William A. Rushworth, *The Sheep* (1899)

beads, and copper for food supplies. A length of copper wire measured from a sheep's head to the tip of its tail was the medium of exchange for the sheep.

Eighteenth- and nineteenth-century slave traders picked up sheep in sub-Saharan Africa and took them to the Caribbean, Central America, and northern South America, where they were mixed with the churro sheep imported from Spain. Later breeders introduced British breeds, particularly Wiltshire Horns, to develop the genetic traits displayed in today's tropical hair sheep breeds.

Sheep played an important role in Scandinavian history, too. Despite the fact that neither sheep nor goats are native to Scandinavia, archaeologists excavating an ancient Danish settlement unearthed a sheep tooth dating to 3980–3810 BCE, proving sheep's presence in Denmark at that early date. Sheep bones dating to the Neolithic period were discovered in Swedish caves. During the Early Iron Age,

Norwegian mourners furnished their dead with mutton for their journey to the afterlife. The Sami of Swedish Lapland sacrificed sheep at ritual sites well into the sixteenth century. Viking wove sails of sheep wool and sometimes carried sheep as a living food supply.

But nowhere in Scandinavia were sheep as important as they were in bleak, newly settled Iceland, where from the time of settlement in the late ninth century through 1900, life depended on sturdy Northern sheep for food, tallow, hides, and wool (during the eighteenth century an estimated 22 percent of the population engaged in working with wool for a full seven months of each year).

### Sheep in Asia

Sheep sustained life halfway around the world in Asia, too. In 1998, Chinese archaeologists unearthed 3,500-year-old sheep bones inscribed with the oldest known examples of Chinese writing. Sheep are also depicted in prehistoric rock paintings in Qinghai, Ningxia, Xinjiang, Menggu, and Gansu in China, as well as in Inner Mongolia, and they feature prominently in numerous artifacts from China's New Stone Age.

The Chinese writing system is one of the oldest known written languages, with some of the earliest examples more than 4,000 years old. The Chinese writing system uses a logographic system (a series of symbols that represent a complete word or a phrase) consisting of large *kanji* (characters). Many Chinese words incorporate the character for *sheep*, including *mouth-watering*, *beautiful*, and *nice*. In archaic characters, the words *sheep* and *lucky* are interchangeable.

## Baa-ck in Time

Many modern breeds have long, illustrious histories, among them the following:

**COTSWOLD** Pre-Roman conquest

**ICELANDIC** Carried to Iceland by Viking settlers in the ninth and tenth centuries CE

**JACOB** Possibly mentioned in the Bible and definitely raised in Britain for more than 350 years

**KARAKUL** Pictured in Babylonian temple carvings

**NAVAJO-CHURRO** Brought to the New World from Spain circa 1493

**SHETLAND** Taken to the Shetland Islands by Viking settlers

**SOUTH COUNTRY CHEVIOT** First recognized in 1372

**TUNIS** Before Common Era

**WELSH BLACK MOUNTAIN** Pre-Middle Ages

The Chinese zodiac originated at least 2,000 years ago; one of the 12 signs of the Chinese zodiac is the sheep. The Year of the Sheep, sometimes called the Year of the Goat (either is correct), occurs every 12 years, rotating with the other 11 signs of the zodiac. The Japanese, Koreans, and Vietnamese all borrowed the Chinese zodiac and use it to name the years.

The current Japanese writing system traces its history to the fourth century, when Chinese characters were introduced to Japan. The character for *sheep* is based on the shape of a sheep's head with two horns, four legs, and a tail. An ancient Japanese saying still heard today is *youtou-kuniku*, meaning "sheep's head–dog's meat," referring to the use of a better name to sell inferior goods.

All of these historic nuggets are the barest tip of the iceberg! The history of sheep is interwoven with that of man. We'll talk more about that bond throughout this book. But for now, let's take a peek at what makes sheep tick.

# CHAPTER 2

# Introducing the Sheep

*'Now take a sheep', the Sergeant said. 'What is a sheep [but] millions of little bits of sheepness whirling around and doing intricate convolutions inside the sheep? What else is it but that?'*

— Flann O'Brien, *The Third Policeman* (1967)

Sheep are reputed to be airheads, but in fact, they are reactive, not stupid. When you're small, tasty, and defenseless, it's best to run now and ask questions later. That doesn't mean sheep aren't smart. As Mark Twain put it, "It is just like man's vanity and impertinence to call an animal dumb because it is dumb to his dull perceptions."

## PARTS OF A SHEEP

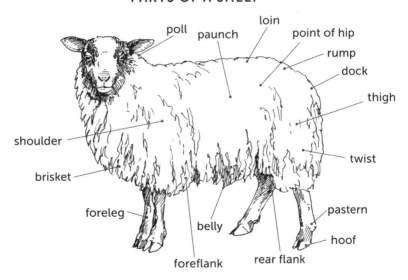

- poll
- paunch
- loin
- point of hip
- rump
- dock
- thigh
- shoulder
- brisket
- twist
- foreleg
- belly
- pastern
- hoof
- foreflank
- rear flank

British researchers, headed by neuroscientist Dr. Keith Kendrick at Cambridge University's Babraham Institute, taught 20 sheep to recognize pictures of other sheep's faces. They showed the sheep 25 pairs of similar faces, training them to associate recognition of the faces with a food reward. Two years later the team again showed the sheep the series of pictures, measuring their brain activity via electrodes implanted in their brains. Some remembered all 50 of the original faces, even in profile. "It's a very sophisticated memory system," explains Dr. Kendrick. "They are showing similar abilities in many ways to humans." Pretty smart for a "dumb animal."

Three years after their first experiment, Dr. Kendrick and his colleagues discovered that sheep prefer smiling or relaxed faces over those of angry humans or stressed-out sheep. They presented the test flock with two doors they could push open to grab a snack. One was covered by a picture of a smiling human or a contented sheep; the other, an angry human or a distressed sheep. They invariably chose the happy-face doors.

Caroline Lee, a member of the animal welfare team at the FD McMaster Laboratory in New South Wales, Australia, developed a test to measure intelligence and learning in sheep based on a complex, 59- by 26-foot (18 x 8 m) maze similar to those used for rats and mice. By penning part of the herd in sight at the exit, researchers used Merino sheep's strong flocking instinct to motivate test sheep to go through the maze. The premise: the time it initially takes an animal to rejoin its flock indicates intelligence, while improvement in times over consecutive days of testing measures learning and memory. It took the 60 sheep an average of 2 minutes to go through the maze on day one and only 30 seconds on day three. When the sheep were retested six weeks later, they went through the maze even faster than they did on day three.

Finally, consider the smart sheep who live on a commons near Marsden, a village on the edge of the Pennine Hills in West Yorkshire, England. Because sheep had been continually raiding village gardens, authorities installed an 8-foot (2.4 m) cattle guard to keep the sheep out of town. Not to be thwarted, the sheep began lying down and rolling, commando-style, across the cattle grid. When BBC News asked a National Sheep Association spokeswoman about their behavior, she replied, "Sheep are quite intelligent creatures and have more brainpower than people are willing to give them credit for."

## The Five Senses

To understand why sheep do the things they do, try to imagine the world through a sheep's five senses: vision, taste, hearing, smell, and touch.

### Vision

To the side, sheep have a wide monocular vision ranging from 270 to 340 degrees, depending on the shape of their faces and how much wool they have around their eyes that may obstruct their vision. Sheep have large, rectangular-shaped pupils, and their eyeballs are placed toward the sides of their heads, giving them a narrower field of binocular vision.

A sheep has a blind spot directly behind her, but by raising or lowering her head or by moving it from side to side, she can scan her

## FIELD OF VISION

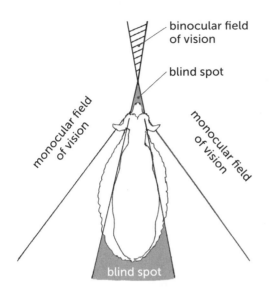

- binocular field of vision
- blind spot
- monocular field of vision
- monocular field of vision
- blind spot

entire surroundings. The rectangular pupils provide a wide-angle effect, giving sheep excellent peripheral vision. They can't easily see objects high above their heads, however, and they have poor depth perception. Because of this, sheep avoid shadows and harsh contrasts of light and dark. They also tend to move away from darkness and toward light.

Based on research that determined the number of cones and rods in sheep's eyes and on subsequent studies to understand how that number related to visual perception, scientists believe that sheep's vision is very keen. They've also learned that sheep see in color, although their color acuity is lower than humans'.

### Taste

Ewes lick their fluid-drenched newborn lambs and this creates a strong maternal bond. Rams sample urine from ewes in heat to determine if they're ready to breed.

Sheep prefer certain feeds over others, indicating that they know what they like to eat. When grazing alongside other species such as goats and cattle, each species selects different types of plants.

### Hearing

Sheep hear well; they are sensitive to high-pitched and sudden noises, both of which trigger a surge of stress-related hormones. They refine their hearing by moving their ears, heads, or entire body to face whatever they're focusing on.

Sheep have scant vocabularies as compared with, say, pigs or cows. Frantic, high-pitched *baas* are stress indicators. Adult sheep give medium-pitched *baas* when seeking older lambs, friends, or feed. Ewes murmur sweet "flutter *baas*" to their newborn lambs. Rams rumble when sexually aroused or annoyed at another sheep.

**LEARN FROM THE PAST**

Many an ignorant flockholder catches and takes hold of the sheep by the wool, at any place he can get hold of best. Men who do this do not realize that the skin of the sheep is very lightly attached to its flesh, and that by holding the sheep by the wool in this careless manner the skin is torn loose from the flesh as far and a little farther than the hand's reach, thus injuring the innocent sheep. It has been our experience that it takes the sheep about two months to recover from the bruise thus caused.

— Frank Kleinheinz, *Sheep Management: A Handbook for the Shepherd and Student* (1912)

## Smell

Sheep have a well-developed sense of smell. Ewes recognize their newborns by taste and scent. Rams sniff ewes to determine which ones are in heat. Sheep are capable of scenting water from afar and they sniff feed to determine if it's safe to eat.

Both sexes also flehmen, whereby the sheep opens his mouth and curls back his upper lip to transfer scent to a structure called the *Jacobson's organ*, or *vomeronasal organ*, located in the roof of his mouth. Rams frequently flehmen when sniffing the urine of a ewe in heat.

### Sheep Tipping

Once a sheep's feet are off the ground and she can't touch anything with her hooves, she remains still. That's why shearers tip sheep to immobilize them for shearing. Tipping also works for routine tasks such as hoof trimming and doctoring minor wounds. Tipping a sheep seems daunting at first but is easy once you know how.

Standing on the left side of the sheep, reach across with your right hand and grasp her right rear flank by the skin (don't pull her wool). With your left hand, bend her head away from you, back against her right shoulder. Lift the flank and pull the sheep toward you. This puts her off balance and theoretically she rolls gently toward you onto the ground. In practice, it doesn't always happen that way. Persevere. Practice makes (almost) perfect.

Hold the sheep on her side, then quickly grasp both front legs and set her up on her rump so she's slightly off center, resting on one hip. If she struggles, place one hand on her chest for support and inch backward until she's more comfortable.

Don't tip a sheep right after she has eaten, as the position puts considerable strain on a full rumen. For the same reason, don't tip pregnant ewes for more than a few minutes at a time. Despite their good intentions, don't let anyone try to help by holding the sheep's legs; if her hooves encounter anything, even your helper's hands, she will struggle to break free.

## Touch

Sheep are less responsive to touch than other livestock species because they are covered with wool or thick, coarse hair. However, sheep have very sensitive skin under their wool. Please, don't pull it!

## Open Your Mouth and Say *Baah*

Estimating a sheep's age by examining the eight teeth in the front of his mouth is easy. If all the teeth are sharp and small, they are baby teeth and the sheep is less than one year old. If the center two teeth are big (these are permanent teeth) and the rest are small, the sheep is between one and two years old. If the four center teeth are big and the rest are small, the sheep is about two years old. If the six center teeth are big and the rest are small, the sheep is about three years old. If all eight teeth are big, the sheep is about four years old. After that the teeth gradually spread and eventually fall out in old age.

A sheep's lower teeth should meet flush with the upper dental pad. If the lower teeth extend beyond the dental pad, she is "sow-mouthed" or "monkey-mouthed"; this is more common in Roman-nosed (arch-faced) breeds. If the dental pad extends beyond the lower teeth, she is "parrot-mouthed."

Extremely parrot-mouthed or sow-mouthed sheep have problems grasping forage when browsing and grazing. They tend to lose weight on pasture and need hay and possibly grain to survive. Both conditions are hereditary and should be avoided; that is, don't breed sheep who display these conformation flaws.

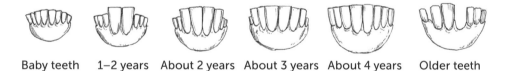

| Baby teeth | 1–2 years | About 2 years | About 3 years | About 4 years | Older teeth |

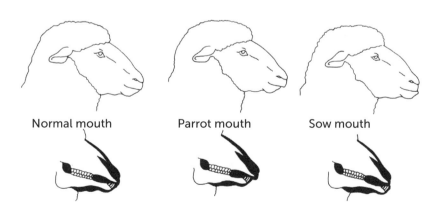

Normal mouth      Parrot mouth      Sow mouth

## Social Behavior

A well-defined pecking order exists within every flock whether composed of four or four hundred sheep. Where a sheep stands in this hierarchy depends on its age, sex, personality, aggressiveness toward other sheep, and the size of (or lack of) its horns. Suckling lambs assume their dam's place in the order and often rank immediately below her after weaning.

Newcomers must fight to establish a place in the flock. Fighting for social position is conducted one on one; established flock members don't gang up on a new one.

When fighting, rams back up and charge one another; both sexes butt, jostle, and push. Other forms of aggression include staring, horn threats (chin down, horns jutting forward), pressing horns or forehead against another sheep, backing up without actually charging another sheep, and ramming an opponent's rear end or side. Very little infighting occurs in a static flock, however, once each member knows and accepts her place.

A major difference between sheep and goats is that sheep don't follow a specific leader as goats do a herd queen. When one sheep starts

## Ewe Said It

### DRINK LIKE A SHEEP?

**10 Sheepy Wineries**

Two Birds and a Sheep Winery (Paso Robles, California)

Inherit the Sheep Winery (Napa, California)

Greedy Sheep Winery (Cowaramup, Western Australia)

Barking Sheep Wines (Mendoza Province, Argentina)

Ram's Gate Winery (Sonoma, California)

Sheep's Back Winery (South Melbourne, Victoria, Australia)

Black Sheep Winery (Murphys, California)

Ram's Leap Wines (Warren, New South Wales, Australia)

Shepherd's Run Winery (Wamboin, New South Wales, Australia)

Black Sheep Vineyard (Adena, Ohio)

**10 Sheepy Alcoholic Beverages**

Barking Sheep Chardonnay Chenin Blanc (Argentina)

Big Horn Buttface Amber Ale (USA)

Red Sheep Oak Aged Shiraz (Australia)

Golden Sheep Pale Ale (England)

Black Sheep Ale (England)

Sheep Head Ale (USA)

Inherit the Sheep Cabernet Sauvignon (USA)

Isaac's Ram Wine (Israel)

Sheep Dip Whiskey (Scotland)

Black Ram Whiskey (Bulgaria)

moving, others follow, even in low-flocking breeds. High-ranking flock members move near the middle of the flock to avoid being in a vulnerable position should predators strike.

It takes a while for sheep to form a cohesive flock when individuals from different sources are mixed. Sheep form friendships and tend to stay in family groups of a ewe and her offspring, their offspring, and so on. In flocks made up of more than one breed, given enough space, sheep generally segregate themselves by breed.

The social group is very important to sheep and isolation is extremely stressful. An isolated sheep may go berserk, bleating loudly, running, and smashing into the sides of the pen. Even in quarantine or a sick bay, sheep should be able to see other sheep. (See chapters 8 and 9 to read about breeding and lambing behaviors.)

## Play Behavior

Lambs begin playing within a few hours of birth. They stay close to their mothers for the first weeks of life, but as they get older, they venture farther and begin to form gangs. By the end of the first month they spend roughly 60 percent of their time with other lambs.

Play behaviors include mounting one another (both ram and ewe lambs do this), playful butting, racing to a set location and back again ("lambpedes"), jumping on and off of rocks or dirt piles en masse, Ninja-style kicking, leaping and whirling in place, and spronking (a Pepé le Pew gait in which lambs race while bouncing stiff-legged, all four feet hitting the ground at the same time). Ram lambs prefer manly games like play-fighting and mounting one another, while ewe lambs

prefer to race. Play becomes infrequent after about nine months of age, although even adult sheep engage in play at times.

## Feeding Behavior

Pastured sheep prefer to camp (sleep and ruminate) on elevated ground, moving downhill to graze during the day. They generally graze in the early morning and again late in the afternoon, resting in their feeding area between grazing sessions. During periods of extreme heat and humidity they often rest from mid-morning to evening and graze in late evening and at night. Sheep generally graze between five and ten hours a day, though grazing time is affected by many factors including weather, breed, quality and availability of pasture and supplemental feed such as tasty hay, and day length. Sheep prefer not to graze near sheep or goat droppings, but they aren't offended by manure from other species.

Sheep have a remarkably dexterous and deeply cleft upper lip that permits very close grazing. As a sheep grazes, she jerks her head slightly forward and up to break grass and stems against the dental pad and lower front teeth. The cleft lip also allows her to sort feed; for example, she may consume all of her favorite bits of yummy corn from mixed grain and leave the oats and barley behind.

Sheep usually ruminate (chew their cud) while resting on their sternums (lying down with one or both front feet tucked under their bodies), but also sometimes while standing. The time spent ruminating is about equal to or slightly less than the time spent grazing. Rumination induces a state of drowsiness that sheep seem to enjoy.

## Flocking, Moving, and Catching Behavior

All sheep flock to some degree, especially when frightened. Flocking instinct, however, varies greatly from breed to breed. Lambs are hardwired to follow their dams and then other members of the flock. Banding together in groups protects sheep from predators that home in on sheep grazing alone or on sheep at the edge of the flock. Flocking instinct is what enables shepherds to tend to and move large numbers of animals.

Fine-wool sheep like Merinos, Rambouillets, and Columbias flock closely while moving, grazing, or at rest; other breeds flock when moving but spread out somewhat to graze and rest. Leading low- and poorly flocking breeds with a bucket of feed is easier than trying to drive them.

Here are some additional points to consider:

**PERSONAL SPACE.** Sheep maintain personal security or flight zones. Anything scary invading a sheep's personal space generates flight. A sheep's flight zone might be 50 feet (15 m) or nothing at all. Breed, gender, tameness, training, and degree of perceived threat enter each equation.

**POINT OF BALANCE.** A sheep's point of balance is at his shoulder, at a 90-degree angle from his spine. Movement within a sheep's flight zone and behind his point of balance makes him move forward; movement in front of his point of balance and within his flight zone makes him turn and move away.

### FLIGHT ZONE

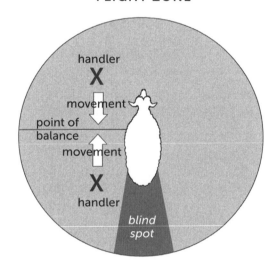

We often see farmers trying to lead a sheep by taking hold of its neck, of course also by the wool, and dragging it along. They make a hard task of it for themselves and certainly make it most unpleasant for the sheep. . . . [To lead a sheep from the left] place your left arm around its neck, and your right hand on the end of its tail-head, tickling it just a little there, and you will find that it will, as a rule, come your way very quickly, at times even faster than you care to have it come.

— Frank Kleinheinz, *Sheep Management: A Handbook for the Shepherd and Student* (1912)

**HERDING SHEEP.** When approaching or driving sheep, don't look directly into their eyes; wolves and herding dogs do that and it instinctively makes sheep nervous. Don't attempt to drive a flock faster than its usual walking speed; rushed sheep tend to scatter.

**MOVING SHEEP.** Calm sheep move forward; frightened sheep move backward. Sheep prefer not to cross water or to move through narrow openings. They move uphill more readily than downhill and prefer to move with the wind rather than against it.

**CATCHING SHEEP.** To catch a sheep, herd the flock into a corner and keep it there by extending your arms or a crook to create a barrier. You can then carefully cut the desired animal out of the group using her flock mates' bodies to prevent her from scampering away. (The smaller the pen, the easier it is to catch individual sheep.) Cornered, frightened sheep will try to squeeze through any small gap to escape and often attempt to jump to safety, usually toward whatever is blocking their way. A cornered sheep can hit an adult at chest height. Be careful!

**RESTRAINING SHEEP.** To temporarily restrain a sheep, place one hand under her jaw and raise her head while steadying her opposite flank or her hindquarters with the other hand. A sheep has considerably more power when her head is down, so keep that head up to maintain control.

**SIGNS OF IRRITATION.** Annoyed sheep stamp their hooves, raise or nod their heads, or glare. They usually back a few paces before charging, although ewes sometimes bash each other from a standstill. Rams or pushy ewes may rub or bump humans with their foreheads. This is a sign of early aggression. Don't allow it. (A sharp smack on the nose should discourage him.)

**BE NICE!** Sheep remember bad experiences for up to two years. You will regret it if you lose your temper.

## Cast Sheep

A sheep who has rolled over onto her back is called a *cast* sheep. She can't get up without assistance. Short-legged, widely built sheep with full fleeces and heavily pregnant ewes are especially prone to becoming cast, as are old, arthritic sheep. The heavy rumen of a cast sheep rests on her lungs, so cast sheep die of suffocation within an hour or so if they aren't helped back up. Roll a cast sheep over so she's propped up on her sternum, then when she's ready, help her up, supporting her until she's steady on her feet.

*A cast sheep cannot right herself and needs immediate assistance.*

## Behavioral Tidbits

Finally, here are a few additional facts to help you understand ovine behavior.

- Lambs are hardwired to move toward any large moving object. This response is strongest between four and ten days of age and explains why lambs approach humans so readily during that period.
- Very young lambs who become separated from their mothers try to return to the spot where they first became separated.
- Healthy lambs sleep about eight to ten hours a day. They seek their dams at naptime and sleep as close to her as possible.
- Older lambs nurse so aggressively that they hoist their dams' hindquarters off the ground. This behavior can lead to udder problems such as mastitis (see page 102) and signals that this lamb should be weaned.
- Sheep look up at their humans' faces. As Dr. Kendrick's experiments prove, they are capable of recognizing many faces, including their shepherds' (see page 14). They are less able to distinguish faces that are looking down or away, or with the eyes covered.
- Humans appear less threatening when they are side-on to a sheep or down nearer the sheep's height. That's why it pays to sit on or close to the ground when taming sheep.
- Sheep are attracted to the sight of other sheep — not only real sheep but also models, mirrors, and photographs.
- Some sheep show a strong lateral preference and consistently move to one side instead of the other, given a choice.
- Healthy sheep and lambs stretch when arising. Immediately after stretching, ewes urinate. When one sheep in a group urinates, the rest usually do, too.

# Choosing a Breed

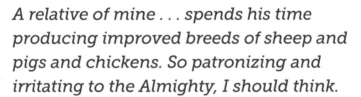

*A relative of mine . . . spends his time producing improved breeds of sheep and pigs and chickens. So patronizing and irritating to the Almighty, I should think.*

— Saki, *The Unbearable Bassington* (1912)

According to some sources, there are more than 1,000 breeds of sheep around the world. At least 75 breeds are available in North America alone. We can't discuss them all in a book this size, but have included a few of the more common or more intriguing breeds. You can find out more by contacting specific breed associations (most of them have websites) or browsing through other sheep books.

## What Are You Looking For?

Make a list of what you want from your sheep before choosing a breed. Do you need a certain kind of wool for handspinning, felting, or weaving rugs? Sturdy sheep for brush control? Parasite-resistant sheep to raise organically? Horned or hornless sheep? Miniature sheep? Heat-tolerant sheep for the South? Club lambs

for 4-H or FFA? Meat sheep? Rare sheep? Heritage sheep? Registered sheep or crossbreds?

If you're looking at registered breeds, find out what the registration policies are for the breed you're interested in before you buy. Most have closed flock books, meaning that to be registered, a lamb must have two parents already registered with that organization. But some registries maintain separate flock books for *percentage sheep*, with ancestors of another breed. Usually, the higher the percentage, the more valuable the sheep.

### Crossbreds and Mules

Some sheep are crossbreds; that is, their parents are of two different breeds. Crosses are often bigger, hardier, and more productive than either parent because crossing results in hybrid vigor, also called *heterosis*. Crossbreeding is used extensively in the commercial

These lists, compiled from many sources including input from my Hobby Farms Sheep Yahoo group, are just a brief overview. Keep in mind that not every individual of any breed necessarily performs up to par.

**Versatile Wool**
(sweaters, hats, socks, soft blankets)
- Bluefaced Leicester
- California Red
- Cheviot
- Columbia
- Coopworth
- Corriedale
- Finnsheep
- Gotland
- Montadale
- Navajo-Churro
- Oxford
- Panama
- Polypay
- Santa Cruz
- Shetland
- Soay
- Targhee
- Tunis

**Fine Wool** (extra soft, clothing)
- Cormo
- Debouillet
- Delaine Merino
- Est à Laine Merino
- Rambouillet

**Sturdy Wool** (heavy sweaters, thick blankets, rugs)
- Black Welsh Mountain
- Border Leicester
- Clun Forest
- Dorset
- Hampshire
- Horned Dorset
- Karakul
- Navajo-Churro
- Suffolk

**Brush Control**
- American Blackbelly
- Barbados Blackbelly
- Florida Cracker
- Gulf Coast Native
- Hog Island
- Katahdin
- Royal White
- St. Croix

**Dairy**

Specialized dairy breeds
- British Milk Sheep
- East Friesian
- Lacaune

Other dairy breeds
- Dorset
- Dorset Horn
- Icelandic
- Katahdin
- Polled Dorset
- Polypay
- Rambouillet

**Parasite Resistance**
- American Blackbelly
- Barbados Blackbelly
- Florida Cracker
- Gulf Coast Native
- Hog Island
- Katahdin
- Royal White
- St. Croix

**Dual-Purpose** (meat and wool)
- Babydoll Southdown
- Black Welsh Mountain
- Bluefaced Leicester
- Border Leicester
- British Milk Sheep
- California Red
- Canadian Arcott
- Charollais
- Cheviot
- Clun Forest
- Columbia
- Coopworth
- Corriedale
- Dorset Horn
- East Friesian
- Est à Laine Merino
- Gotland
- Gulf Coast Native
- Hampshire
- Hog Island
- Icelandic
- Île de France
- Lincoln
- Miniature Cheviot
- Navajo-Churro
- North Country Cheviot
- Outouais Arcott
- Oxford
- Panama
- Perendale
- Polled Dorset
- Polypay
- Rideau Arcott
- Romeldale
- Romney
- Scottish Blackface
- Shropshire
- Southdown
- Suffolk
- Targhee
- Texel
- Tunis

**Club Lambs** (for 4-H and FFA)
- Columbia
- Hampshire
- Southdown
- Suffolk

**Heat Tolerance**
- Dorper
- Florida Cracker
- Gulf Coast Native
- Hog Island
- Katahdin
- Royal White
- St. Croix

**Triple-Purpose** (dairy, meat, wool)
- East Friesian

**Meat**
- Dorper
- Katahdin
- Royal White
- St. Croix
- Texel
- Wiltshire Horn

**Miniatures**
- Babydoll Southdown
- Classic Cheviot
- Miniature Cheviot
- Ouessant
- Shetland
- Soay

sheep industry, and most slaughter lambs are crossbreds. Sheep are sometimes intentionally crossbred to increase a specific trait in their offspring. For instance, crossing to highly prolific breeds such as Finnsheep (Finnish Landrace) result in ewe lambs that will produce more lambs themselves when they're mature.

Mules are a specific type of crossbred especially popular in Great Britain. To produce mules, rams of a specific breed are bred to ewes of another specific breed so that, for instance, Bluefaced Leicester rams are bred to Swaledale and Scottish Blackface ewes to produce North of England mules, while Teeswater and Wensleydale rams are bred to hill breed ewes such as Dalesbreds and Swaledales to produce a wildly popular mule called the *Masham*. Mule ewes are then bred to meaty Suffolk or Texel rams to produce fast-growing market lambs. This is known as the *stratified three-tier system*.

In 2008, the North American Mule Sheep Society brought the stratified three-tier system to North America. It registers mules of various crosses and hosts an annual mule sheep show at the Wisconsin Sheep and Wool Festival in Jefferson, Wisconsin.

## Breed Classifications

Any number of ways are used to classify sheep. I prefer the one Deborah Robson and Carol Ekarius use in *The Fleece and Fiber Sourcebook*, my favorite wool and sheep-breed guidebook. They divide breeds into the following groups, which are described in greater detail starting on page 27.

> Blackfaced Mountain Family
> Cheviot Family
> Dorset Group
> Down Family
> English Longwool Family
> Feral Group
> Merino Family
> Northern European Short-Tailed Family
> Welsh Hill and Mountain Family
> Other Sheep Breeds

Other classification systems include:

**BRITISH WOOL CLASSIFICATIONS.** This is the system used by the British Wool Marketing Board System. It divides native sheep into Mountain and Hill (all breeds of Cheviots, Rough Fell, Herdwick, and so on); Long wool and Lustre (Devon and Cornwall Longwools, Bluefaced Leicesters, Cotswolds, and the like); Short wool and Down (Suffolks, Southdowns, and Shropshires to name a few); and Medium (Jacobs, Border Leicesters, Romneys, and so forth).

**FIBER TYPE.** Fine wool, long wool, medium wool, carpet wool, and hair sheep.

**HERITAGE BREEDS.** These are rare, old breeds promoted by organizations such as the American Livestock Breeds Conservancy, Rare Breeds Canada, and the Rare Breeds Survival Trust of the United Kingdom.

**MINIATURE BREEDS.** Small sheep, generally no more than 24 inches (61 cm) tall, weighing less than 150 pounds (68 kg).

**PROLIFIC BREEDS.** Most sheep have single lambs or twins; triplets are fairly common. Prolific breeds such as Finnsheep, British Milk Sheep, and sheep carrying the Booroola Merino gene, however, have litters, sometimes up to seven lambs per litter.

**PURPOSE.** Meat, wool, dairy.

**SIRE OR DAM BREEDS.** Sire breeds like Suffolks are strong in meat-making characteristics, while dam breeds excel in reproductive traits.

**TAIL TYPE.** Fat-tailed sheep deposit fat in their rumps or tails instead of internally like other sheep; examples include the Damara and Karakul. Northern European short-tailed breeds like Shetlands, Soay, Finnsheep, and Icelandics have short, fluke-shaped tails.

**WHITE-FACED VS. BLACK-FACED BREEDS.** White-faced sheep such as Merinos, Rambouillet, and Columbias are generally wool sheep, whereas black-faced sheep (which should actually be called colored-faced sheep) such as Hampshires, Suffolks, and Southdowns are noted for meat.

## What Are All Those Numbers in Breed Descriptions?

Until the late 1970s, wool was evaluated by the Bradford count, also called the *spinning count*, a system by which experts estimated how many 560-yard (512 m) hanks of single-strand yarn could be made by a good spinner from a pound of cleaned, combed wool with the fibers all parallel. The finer the average diameter of a single wool fiber, the more hanks that could be spun. From a pound of 44s, for example, 44 hanks could be made. More than 80 hanks could be spun from the finest fleece; from the heaviest, 36 or fewer. Though more objective measuring systems are rapidly replacing the Bradford count in the international market, it is still used among some shepherds and breed associations.

Nowadays fleece is usually measured in microns. Small samples are measured in the lab using a Sirolan Laserscan or an Optical Fiber Diameter Analyzer, or in the field using a mobile instrument called a *Fleecescan*. A micron is a measure of diameter equal to one-millionth of a meter, or 1/25,000 of an inch. What is usually referred to as a fleece's micron count is the average diameter established by analyzing samples from several sites on each sheep. Most wool falls into the 18 to 40 micron range.

Fine wool suitable for next-to-the-skin clothing measures in the 18 to 24 micron range. Wool in the 25 to 38 micron range is usually used for knitwear apparel and woven into blankets; most handspinning fiber falls into this category. Wool testing 38 microns and greater is generally utilized by the carpet industry, but handcrafters also favor certain types to make specialty items such as heavy woven rugs and saddle blankets, and for felting.

Several laboratories (see Resources) micron-test fleece in the United States and many more are located overseas.

## BLACK-FACED MOUNTAIN FAMILY

The black-faced mountain breeds originate in the hill country of Wales, Scotland, and northernmost England. They are amazingly hardy sheep, able to thrive on rough grazing and to withstand cold, wet, and bleak conditions, yet still produce a full fleece of wool and tasty lamb or mutton.

The only breed in this group available in North America is the intrepid Scottish Blackface. Roughly 30 percent of all the sheep in England, Scotland, Wales, Ireland, and Northern Ireland are Scottish Blackface, but they're a rare breed throughout North America. Like others in the black-faced mountain group, Scottish Blackface grow long, coarse, resilient carpet wool that is also used for stuffing mattresses. Black-faced mountain breeds available in Britain include Dalesbreds, Derbyshire Gritstones, Lonks, Rough Fells, and Swaledales.

## CHEVIOT FAMILY

Cheviots hail from the bleak Cheviot Hills on the border between Scotland and England. Cheviots' faces and legs are free of wool. Like other British hill and mountain breeds, Cheviots are good grazers and extremely hardy. They are active, agile, perky sheep with upright, horselike ears and medium-grade wool. Some, like the Classic and Miniature Cheviots of North America and the Brecknock Hill Cheviot of Wales, come in colors as well as white. Cheviots are dual-purpose sheep well known in Britain for their fine eating qualities.

Cheviot breeds available in North America include the commercial Cheviot, Classic Cheviot, Miniature Cheviot, and the North Country Cheviot. Additional British breeds include the South Country Cheviot (formerly called the Border Cheviot), the Brecknock Hill Cheviot of Wales, and the Wicklow Cheviot of Ireland.

### DID EWE KNOW?

**Brockle-faced** is a black-and-white splotched facial pattern seen on crosses between white-faced and black-faced breeds; also called smut-faced.

Dorset: polled ewe and horned ram

Babydoll Southdown

## DORSET GROUP

In North America, the Dorset group consists of Polled Dorsets and Horned Dorsets; Britain has their distant cousins, Dorset Downs, as well. Polled Dorsets are a recent development of the Dorset Horn (also called the Horned Dorset). Dorset Horns are an ancient breed and once very popular worldwide, though now they're listed as threatened on the American Livestock Breeds Conservancy's conservation priority list. They were widely used in developing many other breeds including the popular Dorper meat sheep. One reason for their popularity was that they breed out of season and are capable of lambing three times in two years. They are good meat sheep, have lovely fleece, are also very milky, meaning they produce a lot of milk for their lambs. Apart from specialized dairy sheep, they're best bets for home dairying.

## DOWN FAMILY

The Down breeds were developed primarily as meat sheep in the chalky downs region of southern England. They have short wool, strong bodies, and colored faces. Most are white but colored strains occur in some breeds.

They are large sheep (with the exception of the old-style Babydoll Southdown), fast maturing, and prolific. Most 4-H and FFA club lambs are Suffolks or Hampshires. Other down breeds include Southdowns (commercial-size and Babydolls), Oxfords, Shropshires, and in Britain, Dorset Downs.

## ENGLISH LONGWOOL FAMILY

The "Blakewell or Leicester" is now called the Leicester Longwool (or the English Leicester in some parts of the world); the Lincolnshire is the Lincoln; and the Romney-Marsh, today's Romney. Kentish and Brampton longwools are now extinct.

Lincoln

Long-wool breeds available in North America include the Leicester Longwool, Bluefaced Leicester, Border Leicester, Lincoln, Romney, Cotswold, and upgraded American versions of Teeswaters and Wensleydales. Additional British long wools include Greyface and Whiteface Dartmoors and the Devon and Cornwall Longwool. All long-wool breeds are large sheep that grow fleece that hangs in long, strong, wavy-to-curly, lustrous locks. Many are good meat sheep as well. The Lincoln is the heaviest breed of sheep available in North America with rams weighing up to 350 pounds (159 kg).

## FERAL GROUP

Gulf Coast Native

Feral sheep are domestic sheep gone wild. They're common in some parts of the world; New Zealand alone boasts 13 breeds of feral sheep. Three re-tamed North American feral breeds include Gulf Coast Native, Hog Island, and Santa Cruz sheep.

## MERINO FAMILY

Merinos are famous for their ultrasoft wool. Spanish breeders developed the parent breed, the Spanish Merino, during the twelfth and thirteenth centuries. On pain of death, none

Merino

were exported until the sixteenth century, when Spanish monarchs began giving gifts of sheep to their royal relatives in other European countries. Today dozens of Merino breeds have spread around the world, including the fabulously productive Merinos of Australia. Merino breeds available in North America include the Delaine Merino, Est à Laine Merino, Debouillet, and Rambouillet.

## NORTHERN EUROPEAN SHORT-TAILED FAMILY

Shetland

The Northern European short-tailed breeds hail from Scandinavia, various islands scattered across the North Atlantic Ocean, and even northern Russia. A trait they share is a tendency to double-coatedness (double-coated breeds have a soft, woolly inner coat covered by a longer, coarser outer layer of somewhat hairy wool) and all have naturally short, fluke-shaped tails. Two northern European short-tailed breeds, Shetlands and Icelandics, are very popular with North American shepherds and handspinners. Another rarer member of this group, the tiny, formerly feral Soay is gaining in popularity, too.

Additional Northern European short-tailed breeds available in North America include plushy American Gotlands, teensy American Ouessants, and ultra-prolific Finnsheep and Romanovs who give birth to three to six lambs per litter.

Intriguing British breeds of northern European short-tailed extraction are fine-wooled, feral Boreray sheep; extremely rare Castle-milk Moorits; seaweed-eating North Ronald-says, and multi-horned Hebrideans and Manx Loaghtans.

Black Welsh Mountain

## WELSH HILL AND MOUNTAIN FAMILY

Only one of several Welsh mountain sheep breeds is available in North America: the true-black Black Welsh Mountain sheep. Elsewhere, other breeds include Badger-Faced Welsh Mountain, Balwen Welsh Mountain, Beulah Speckled Face, Hill Radnor, Kerry Hill, Llan-wenog, Lleyn, Welsh Mountain and South Welsh Mountain, and Welsh Hill Speckled Face sheep.

Jacob, Texel, Karakul, and Navajo-Churro

## Other Sheep Breeds

The rest of sheepdom as represented in *The Fleece and Fiber Sourcebook* falls under the heading of Other Sheep Breeds, and dozens are listed, ranging from the cinnamon-tinted Tunis to meaty Charollais to multi-horned Jacobs and Navajo-Churros.

In addition to the wool breeds mentioned above, there are a number of sheep breeds that have hair instead of wool.

## Hair Sheep

The first domestic sheep weren't woolly. Some sheep still aren't. According to Susan Schoenian, Extension Sheep and Goat Specialist at the University of Maryland's Western Maryland Research and Education Center, approximately 10 percent of the world's sheep are hair sheep. An estimated 90 percent are found in Africa and 10 percent in Latin America and the Caribbean. We have some in North America, too.

Meat raisers are increasingly turning to ultra-meaty hair sheep such as the Dorpers and White Dorpers of South Africa. These breeds, along with the all-American Katahdin and rare British Wiltshire Horn, are "shedders"; they grow a modicum of wool for winter protection and spontaneously shed it around midsummer.

Latin American and Caribbean hair sheep breeds such as Barbados Blackbellies, St. Croix, and Royal Whites never grow wool at all. Their strong hooves and natural heat- and parasite-resistance make them favorite sheep for the American Southeast, where the humid climate encourages parasite growth.

A third group of hair sheep is registered by the United Horned Hair Sheep Association. These include strongly horned Painted Deserts, Texas Dalls, Desert Sands, Black Hawaiians, Corsicans, and Mouflon. While some are raised for meat and a few for zoos, most rams are marketed to stocked hunting reserves.

Barbados Blackbelly, Katahdin, and Painted Desert

# Horns 101

Before we close, I'd like to say a few words about breeds with horns. Horns are beautiful and functional but have their downsides, too. Sheep come in many horn configurations: polled (meaning they're born, and remain, hornless), sex-linked (rams usually have horns, ewes generally do not), two-horned (both sexes are horned, rams more strongly than ewes), polycerate horned (both sexes have at least two horns, some have multiples numbering up to six), and scurred (rudimentary horns that look like smallish knobs instead of true horns).

A sheep's *phenotype* is his looks — what you see when you observe him. A sheep's *genotype* is his genetic makeup. Because the ability to grow horns is a recessive trait, some seemingly polled sheep are genetically capable of producing horned lambs when mated with horned sheep (or another seemingly polled one carrying the gene for horns). In other words, what you see isn't always what you get. The only way to be absolutely certain your lambs won't be horned is if both parents are purebreds of a polled breed.

On the one hand, horns, be they of sheep, goats, cattle, or any other horned being, dissipate heat, so horned sheep are better adapted to high heat and humidity than are hornless sheep. And, although they hate it, adults' horns make handy handles for snagging and dragging sheep in emergency situations.

On the other hand, horned sheep, especially lambs and small breeds, catch their horns in gates, feeders, and fences; because of this, Electronet-type fencing should never be used with horned sheep. Once entangled in it they cannot get away, receive jolt after jolt of electricity, and sometimes go into shock and die.

Horned sheep, especially rams, love to hear their horns bang on fences, barn walls, and just about anything else they can batter. Expect lots of fence and building maintenance if you keep horned rams.

And even though battling rams tend to connect at the poll instead of with their horns when they charge, horned rams often injure polled rams. If you have both kinds they'll need separate quarters.

## What Is a Miniature Sheep?

There are three types of miniature sheep:

- Naturally tiny breeds that evolved as small animals to better survive the conditions nature handed them. Soays, Shetlands, and Ouessants pop to mind.

- Small breeds that retained their original breed character when their parent breeds were selected for greater size. Classic and Miniature Cheviots and Babydoll Southdown sheep are good examples.

- New, true miniature breeds that are deliberately downsized through crossbreeding and selecting for smaller stature: Miniature Jacobs and Miniature Katahdins, for example.

Owners of traditional breeds in the first two groups are quick to point out that their favorite breeds weren't miniaturized by man.

## Handling Horns

Horn management is another issue. Rams' horns spiral back and then up around the sides of their faces. Very often the first spiral grows too close to the face and brushes against it or even grows into the ram's cheek or eye. Sheep chew side to side instead of up and down, so heavy horns pressing tight against both cheeks make it impossible to chew. Some shepherds say if you can insert a finger between horn and eye, that's sufficient. But as a ram grows older, both his face and horns become more massive, closing off that precious, small space. Then you have to trim his horns. This is not a task for the faint of heart.

Horns are living, bony structures that grow upward and outward from the base of the skull. Inside is a network of blood vessels that bleed profusely if a horn is injured. A horn's tough outer covering of keratin is called the *horn sheath*. It will pull off of horned lambs up to one year of age with relative ease, leaving the exposed horn core bleeding and raw. The sheath, however, keeps growing and in a few months' time re-covers the core. The horns of such a lamb will be asymmetrical for the rest of his life.

Horn buds of lambs are proportionate to the size of the lamb. Large ram lambs of strongly horned breeds such as Scottish Blackface and Icelandics are sometimes born with visible horns. These occasionally cause problems at lambing time, so plan to be there to gently ease the ewe's vulva over her lamb's horn buds so they don't damage the ewe. Most horned lambs, however, sprout horns at one to three weeks of age.

### A HORNUCOPIA

Horns come in a variety of configurations. Shown here are (clockwise from left) a horned ewe, a ram with spiral horns, a Jacob ram with multiple horns, and a sheep with scurs.

*Rams often serve as mascots for armies and sports teams. This is Private Derby, a Swaledale mascot of the 2nd Battalion Mercian Regiment in Derby, England.*

Should a sheep break his horn or a lamb lose his horn sheath, apply Blood Stop powder to staunch bleeding. When bleeding stops, carefully clean the injury only if needed (remember, exposed horn tissue is very, very sensitive) and apply a spray-on wound dressing such as Blue Kote or Betadine every day until the surface is dry. If flies are a problem, apply insect repellant at the base of the horn but not on the horn core itself.

## Trimming a Horn

If you have to trim a horn that is spiraling too close to a ram's face, the best tool is a length of obstetric wire on two rings that fit around your middle fingers (get it from your veterinarian and ask her to show you how to use it), but a hacksaw will do. Here's how to proceed:

- Restrain him with a head gate, fitting stand, or milking stand, or ask an assistant to set him up on his rump and hold him for you.
- Slice into the horn surface, parallel to his face, in small increments, watching out for the sensitive core.
- If you reach core before the job is done, stop, apply hemostatic powder (to stop bleeding), and wait a few weeks for the exposed horn to become less sensitive before trying again.

A better bet, in my opinion, is to have your vet trim the horn under local anesthetic.

## Follow Your Heart

*Choose a breed that tickles your fancy, one you really like. It makes a difference. Here's a story of my own dream come true.*

**For 63 years,** "happy birthday" has been an oxymoron for me. No matter how eagerly I anticipate the day, if something can go wrong, it does. Oh, there have been good birthdays, but some have been awful, like the one when my old dog died. This year was different. Margot Alice, a shepherd-friend then living in Johnstown, Ohio, gave me a pair of ewe lambs. My husband, John, agreed to pick them up in time for my birthday in May. Then Lori Olson, my sister-in-sheep from Boscobel, Wisconsin, asked if she could keep me company while John was away. What fun! Lori hadn't visited in years.

A bit of background: In 1977 a British pen pal mailed me a beautifully photo-illustrated book about British sheep breeds. Though a dozen breeds beckoned, one stood out among the rest: Scottish Black-face. Also known as *Scotties* or *Blackies*, they are an ancient carpet wool and meat breed with impressive horns and long, white fleeces set off by white-speckled black faces and legs. I was totally smitten!

However, like many other wonderful British breeds, Scottish Blackface are scarce in North America. So when my dream of keeping sheep came true, I chose old-time Cheviots, another British hill breed. A few years later, I discovered Blackies being raised less than 50 miles from our home. We went, I saw, I seriously lusted for Scottish Blackface.

One thing held me back. Cheviots are polled, while Scottish Blackface rams have long, strong, spiraling horns, and horned and hornless rams don't mix. I couldn't risk my other rams; we have only a single ram paddock, and John hates building fences. Fast-forward to Lori's arrival. John rushes out to greet her while I yank on my shoes. They're grinning as John calls out, "Come see the puppy Lori brought."

"Puppy?" I'm thinking, when John whips open the back door to reveal . . . a Scottish Blackface lamb in an airline crate! "Happy birthday!" they cry in unison. *"Baaah! Baaaaaaah,"* cries the lamb. I am poleaxed for a heartbeat; then I realize that it's a seven-week-old, registered Scottish Blackface ram! John's going to fence a separate pasture for the little guy.

I name the bonnie wee lad Othello. It's my best birthday ever.

# CHAPTER 4

# Buying Your Sheep

*Now I have a sheep and
a cow, everybody bids me
good-morrow.*

— Ben Franklin

Before you buy sheep, make sure you want them. Try to spend some time with sheep, especially at lambing time, before you commit; this is for your sheep's sake as much as your own. Make certain your way of life will mesh with keeping sheep. Keeping livestock is a huge commitment, with daily responsibilities. You can't buy sheep, then turn them out and forget about them. Here are some questions to ponder before purchasing:

- Are there zoning laws or fencing laws that might prevent you from keeping sheep?
- Will nearby neighbors complain?
- Are you able to provide housing and an exercise area or pasture?
- Can you build adequate fencing and predator protection?
- Do you know how and what to feed sheep?

- Are you prepared to provide hoof care and other routine health maintenance for your sheep?
- Is there a veterinarian in your area who is experienced with sheep?

And most important: Sheep are extremely social creatures who need the companionship of other sheep; you can't keep one sheep all alone.

Once you're sure that sheep are for you, decide what you want from your sheep. Will you buy crossbreds (sheep with parents of two breeds) or purebreds; registered stock or not? Will you raise lambs or start with adults? How many do you need? How much money are you willing to spend to buy and maintain them? How far will you travel to visit farms and buy your sheep?

## Before You Bring Your Sheep Home

**You will need:**

- A permit to keep sheep if you live where permits are required

- Safe shelters, bedding, feed, secure fencing, feeders, water containers, and a consistent source of clean drinking water

- Halters, leads, hoof-trimming equipment, and hand or electric sheep shears if you plan to shear your own sheep, as well as specialized equipment (a milking stand, show halters, a fitting stand, and so on) if you need it

- A well-stocked first-aid kit

**And most important:**

- Phone numbers of at least two reliable veterinarians who are familiar with sheep and will see patients after hours and on weekends, and phone numbers or e-mail addresses of one or two shepherds willing to offer a helping hand as problems arise

- If you plan to breed sheep, you will also need safe lambing quarters, material to make jugs (temporary mothering pens), and a well-stocked lambing kit. If you keep your own ram he'll need a companion (another ram or a wether) and strong, secure quarters with stout fences.

## Where to Start Looking

Let's assume you've read chapter 3 and chosen a favorite breed. Let's make it a popular breed such as Suffolk. Where are you going to find them?

The best, most reliable way to learn how to raise sheep in the area where you live is to discuss your needs with your County Extension agent. What works for sheep in Maine isn't necessarily right in Oklahoma, Washington, or Mississippi. Agents are also your best source for reliable information about local feeds, mineral deficiencies, and effective deworming procedures, and their services are free. To locate County Extension offices in your area, visit the USDA Cooperative Extension System website.

### Finding Sheep for Sale

Check for "sheep for sale" notices on bulletin boards at feed stores and veterinarian practices, or pin up a "Suffolk sheep wanted" sign of your own. Read the local classified ads, especially in pennysaver–type shoppers. Talk to veterinarians and County Extension agents in your buying area; they know who is raising sheep near your home.

Take in a sheep show. Visit information booths and chat with exhibitors between classes. All state and most county fairs host sheep shows. Breed associations sanction them, too; e-mail or call organizations for dates and times.

Join sheep-related e-mail groups. Such affiliations are a great way to find sheep and related supplies, and you'll make new friends who just might own the sheep you want to buy.

Visit breeders' websites. Type the name of the breed, "sheep," and "sale" into the search box of your favorite search engine (e.g., *Suffolk sheep sale*). Qualify it further, if you like, by state (e.g., *Suffolk sheep sale Ohio*). If breeders' websites don't offer what you're searching for, e-mail and ask if they have it. It they don't, they may know someone who does.

Check out ads in sheep magazines such as *Sheep!*, *The Banner Sheep Magazine*, and *Sheep Canada* (see Resources).

---

### Beatrix Potter's Sheep

Beatrix Potter (1866–1943), author of *The Tale of Peter Rabbit* and dozens of other children's classics, helped preserve Cumbria's endangered Herdwick sheep when British farmers abandoned them in favor of more popular, improved breeds. When she died, she left 14 farms, 4,000 acres of land, and her large flock of prize-winning Herdwicks to the British National Trust.

Hardy, hill-dwelling Herdwick sheep have been grazing the steep, bleak Lake Country fells since at least the twelfth century CE. Herdwicks have pale blue-gray fleeces and white faces, but their most unusual feature is their "hefting" instinct; they live their lives within a few miles of their unfenced birth places. When sold away from their homes, Herdwicks try desperately to return.

---

If your chosen breed is a rare one, for example, Hog Island or Karakul sheep, log onto the American Livestock Breeds Conservancy website (see Resources) and click on Classifieds in the menu.

Finally, visit online sheep directories such as the United States Sheep Breeders Online Directory and visit registry websites to peruse their online member-breeder directories. Or, phone or e-mail organizations and ask them where to buy their breed of sheep in your area.

### *Where* Not *to Buy Sheep*

The first rule of sheep buying is to buy from individuals, not from sale barns. People send sick sheep, ornery rams, and problem breeders to the sale barn for their salvage value. In most cases you have no contact with the seller. Was the ewe you purchased vaccinated? Bred? If so, to what sort of ram and when is she due? She or her flock mates might have any of several catastrophic, slow-incubating diseases such as caseous lymphadenitis, hoof rot, or ovine progressive pneumonia (see chapter 7), things you definitely don't want on your farm. And the sheer number of sheep and goats (goats carry most of the diseases that sheep do) passing through a sale barn leads to a natural buildup of disease organisms. Sheep who weren't exposed to disease before they were sold through a sale barn are likely to be infected by day's end.

If you do attend such sales, scrub your hands with plenty of soap the moment you get home and sanitize the clothing you wore. Do it before going near your own sheep or other livestock. Use one part household bleach to five parts plain water in a spray bottle to thoroughly spritz boots and shoes, and launder your other clothing in hot water and detergent. Hoof rot, soremouth, respiratory diseases, and caseous lymphadenitis can hitchhike home on your hands, clothing, and especially your shoes.

If you do buy sale barn sheep, quarantine them for at least 30 days.

### Quarantining Incoming Sheep

When you bring sheep to your farm, be they new animals or those returning from a show or from being bred, plan to quarantine them away from your existing flock. House them in an easy-to-sanitize area at least 50 feet (15 m) from any other sheep or goats but within sight of other sheep; worm them, vaccinate them, trim their hooves, and keep them isolated for at least 30 days. Don't forget to sanitize the conveyance you hauled them home in.

During that time, feed and care for your other sheep first, so you can scrub up after handling the quarantined sheep. Never go directly from quarantined animals to your other sheep. If you can prevent it, don't allow dogs, cats, poultry, or other livestock to travel between one group and the other. When their time in quarantine is up, sanitize the isolation area and any equipment you've used on the quarantined sheep.

## Finding a Responsible Seller

If you're buying locally, tap into the local sheep grapevine. Ask who other sheep owners buy from, who they avoid, and why. After you've narrowed the field to a handful of producers selling your type of sheep, contact them and arrange to visit their farms.

Be courteous and arrive on time. If you have sheep or goats at home and the seller wants to sanitize your shoes, don't be offended. In fact, consider biosecurity precautions a plus.

Look around. Sheep farms aren't necessarily showplaces but they shouldn't be trash dumps either. Are the sheep housed in safe, reasonably clean facilities? Are there droppings in the water tanks? Are they eating poor-quality hay? Are any sheep limping? Do they cough? Do you see runny eyes or noses? Are the sheep in good flesh, neither scrawny nor over-fat? In large herds, you'll spot a few sheep who are skinnier or fatter than the norm, but the majority should be in average condition.

Ask about the seller's vaccination and worming philosophies, particularly which vaccines and wormers he uses and why. How often does he vaccinate and deworm his sheep? Does he test for diseases like caseous lymphadenitis and ovine progressive pneumonia, and does he have documentation that he does? Are any of his sheep currently infected? What about hoof rot? Soremouth? If he's had these problems in his herd, what did he do to control them?

Does he show you only the sheep you arranged to see or the entire flock? Try to see them all, especially sheep related to the ones you came to buy. If there are problems, you should know about them up front.

Ask why the sheep are for sale. Is he changing bloodlines? Downsizing? Switching breeds? If they're culls, perhaps the trait he's culling for doesn't matter to you.

If you like what you see, ask to examine the sheep's registration papers and their health, vaccination, worming, and production records.

*The best advice about training a sheep dog is to use trained sheep. "Naïve" sheep, ones that have never been herded, are tough for any but the more seasoned working dogs to handle.*

## Checking the Paperwork

When buying registered sheep, carefully examine the paperwork to make sure you're getting what you pay for. Papers are transferred after every sale, so they should be issued in the seller's name. If they aren't, he can't sign a transfer slip, and the papers can't be transferred to you.

Most registries stipulate that lambs be registered by their breeders. If you buy eligible but as yet unregistered lambs, ask for a fully filled out and signed registration application and a completed transfer slip conveying their ownership to you.

And be aware that unethical or unknowing breeders sometimes sell sheep that aren't quite what they seem. Full-blood (100 percent) Dorpers, for example, are not the same as percentage Dorpers. Given two sheep of equal quality, the full-blood is worth a lot more money. Learn the jargon before you go shopping.

Production records should indicate a sheep's birth status (single, twin, triplet, etc.) and particulars about her reproductive career. Ask about a ewe's lambing habits. Has she had any birthing problems? Has she ever rejected a lamb? In prolific breeds, does she produce enough milk to feed all of her lambs or do they require supplementary bottle-feeding?

Ask about guarantees. Some producers give them, some don't; if any sort of guarantee is extended, get it in writing.

Is the seller willing to work with you after the purchase, should questions or problems arise?

Above all, trust your intuition. If the seller makes you feel uneasy, thank him for his time and shop elsewhere. Given the many honest sellers in the sheep world, why deal with someone you don't quite trust?

## Evaluating Sheep

Choose individual sheep based on your particular needs, taking the following factors into consideration.

### Health

Always buy sound, healthy sheep. Chapter 7 explains how to tell sick sheep from healthy ones. Make no exceptions. If the sheep you have your heart set on has, for instance, foot rot, and you've never had foot rot on your farm, pass her by. Never knowingly buy expensive trouble.

### Conformation and Type

Another important thing to do before buying sheep is to ask the organization that registers your breed of choice for a copy of their breed standard. This is a list of points to look for when evaluating that breed of sheep. Don't omit this step.

### Availability

If you want something rare, even if it's only rare where you live, plan to devote time, energy, and travel to buying sheep. Otherwise, talk to shepherds in your area and see what breeds and types of sheep they raise. If you choose along the same lines you'll never have to travel far for replacement sheep, stud service for your ewes, or knowledgeable advice.

### Gender

We'll talk more about rams in chapter 8 but it bears saying this now: if you don't plan to breed sheep, don't buy a ram. Rams can be sweet and charming, but they're always unpredictable and it takes reinforced housing and fencing to contain them.

If you're looking for pets, brush control, sheep for herding dog training, or lovely fleece-bearing sheep, think wethers. They aren't preoccupied with lambs, heat cycles, or rut, making them pleasant sheep to have around.

### Age

If you're looking for pets, think about starting with bottle lambs. Bottle lambs bond with the humans who raise them and the type of animals they find in their environment.

If you're getting into breeding, start with experienced, middle-aged, bred ewes who can teach you the ropes. Your first lambing season will be scary enough, so don't start with ewe lambs, who are just as confused as you are and more prone to birthing dystocia (lambing problems) than ewes in their prime.

In fact, I recommend starting with older ewes culled from large operations due to age. Many large-scale producers begin culling ewes at about 6 years of age. Ewes should be excused from the lambing pen when they're about 10 years old, as lambing problems increase with age, but a healthy 6-year-old knows the ropes and has several productive years ahead of her. That's how my own flock began, and that's why this book is dedicated to that first remarkable sheep, Baasha.

## Identifying Sheep

Purebred sheep must be permanently identified before registries admit them to their flock books. This is especially so in breeds where individuals tend to look alike.

Think of two white sheep. You can tell them apart ("Hope has black speckles on the tips of her ears and she has a Roman nose, but Jacy's always fatter than Hope and her facial profile is straight"), but what if something happens to you? If your executor just sees two white sheep and can't match them with their registration papers, their pedigrees will be lost. All sheep should be positively identified in some physical way, even if they are just a couple of beloved pets.

In addition, the United States Department of Agriculture's mandatory USDA Scrapie Program stipulates that every sheep who leaves your farm must be permanently identified with a program-approved ear tag. So unless your sheep never leave home, tags are a must.

### Ear Tags

The beauty of ear tags is that they're easy to read from a distance and inexpensive to buy and apply. They come in metal and plastic in a wide range of types and sizes. Their main failing is that the numbers fade with age and the tags themselves are easily lost. For this reason, large-scale sheep producers sometimes tattoo *and* tag.

The USDA scrapie programs for sheep and goats are overseen on the state level, so you may not have a choice in the type of ear tags you use. If you do have a choice, however, avoid button-type plastic tags and metal tags; both tend to cause infections. Instead, choose small, two-piece plastic models that swivel; they're less likely to snag on fencing or brush and tear the ear.

Plastic tags can be purchased preprinted by companies like Premier1 or as blanks for you to fill in. When writing on blank tags, use a marking pen designed specifically for that pur-

Different types of ear tags

pose. Mark the tags the night before tagging to allow them to dry overnight.

Some colors are easier to read than others; when visibility is an issue, in descending order from most visible to least, choose yellow, white, orange, light green, black, pink, purple, gray, brown, red, medium or dark green, then blue.

Use colors and location to gather information at a glance. For instance, tag males in one ear, females in the other. Tag colors can indicate sex, sire, year of birth, commercial versus purebred animals, and so on.

Premier1 also prints individual tags to order; name-bearing ear tags are a very handy thing when farm sitters or others who are unfamiliar with your animals have to look after your sheep.

### INSTALLING EAR TAGS

To install an ear tag, place the sheep in a handling chute, on a fitting or milking stand, or halter and tie her where you can easily access her ears. If possible, recruit an assistant to help restrain the sheep. Clean the ear with alcohol and pat it dry. Place the tag in your tag applicator and apply antiseptic salve to the male (pointed) tip of the ear tag. Place the female

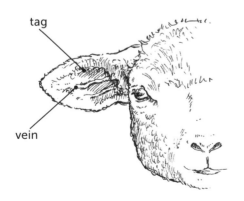

Tag placement on ear

side of the tag inside of the ear and the male part outside. When setting the tag, avoid blood vessels, ridges of cartilage, and any existing scar tissue, and insert it 1 to 2 inches (2.5 to 5 cm) from the skull where ear tissue is thicker and more difficult to tear. Press quickly and firmly, and it's in. Keep an eye on the tag for a few weeks in case infection sets it (although it rarely does when using two-piece, swivel tags).

### Tattoos

Sheep registered with some registries, such as the Icelandic Sheep Breeders of North America, must be ear-tattooed. Even when it isn't required, tattooing is a good way to permanently identify your animals. If specific requirements are given, such as putting a scrapie tag in one ear and a herd identification tag in the other, make sure you put them in the correct ear; that is, the animal's left and right, not yours as you are facing her.

Tattoo pliers come in .300, $5/16$-inch, and $3/8$-inch digit sizes. If you have a five-digit herd prefix (get this from your registry if you're tattooing to their standards), make sure the set you choose accepts five digits (some don't). Better pliers have a positive ear-release feature that pops back open after you've tattooed an ear.

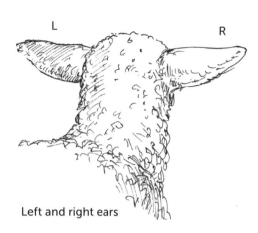

L    R

Left and right ears

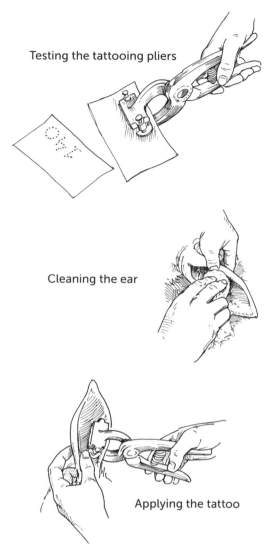

Testing the tattooing pliers

Cleaning the ear

Applying the tattoo

Most tattoo kits come with a roll-on applicator of black ink, but throw it away and buy green paste ink instead. Green is better because it shows up well in dark ears, and paste is better because the roll-on kind tends to drip.

You'll also need a soft toothbrush to scrub ink into newly applied tattoo piercings, disposable gloves, and alcohol and cotton balls or paper towels to clean the site before tattooing.

### APPLYING TATTOOS

1. To tattoo a sheep, assemble your tools before you begin. Sterilize the tattoo digits by immersing them in alcohol and then place them in the pliers.
2. Test the tattoo on paper (index cards work well) to make certain you've inserted the digits right side up and facing the right direction.
3. Restrain the sheep using the same protocol used for ear tagging. A little feed will help distract most sheep, at least until you clamp the pliers on his ear.
4. Swab the inside of the ear with alcohol. Pat it dry.

5. Have a helper securely steady the animal's head. If you aren't using a chute or stand, ask a second helper to crowd the animal against a wall to help secure him in place.
6. Flatten the ear as best you can. Position the pliers in the ear, making certain the digits are inside the ear and you aren't holding

the pliers upside down. Avoid tattooing into blood vessels, ridges of cartilage, or scar tissue. Press quickly but firmly and then release. If the needles stick to the ear or pierce through to the other side, peel the ear away from the needles and apply less pressure the next time.

If some of the punctures bleed, pinch a paper towel or cotton ball over the bleeder until it stops. Don't apply ink until the bleeding stops because blood flow washes ink out of the holes.

7. Squirt a blob of ink onto a soft toothbrush or your gloved finger (tattoo ink is hard to get off, so wear gloves), making certain you get pigment out of the tube, not just oil. Rub the ink into the piercings, taking care to get ink into every hole. When they're filled, leave the excess ink on the ear (scrubbing it off might dilute ink in the piercings as well); it will flake off once it dries.

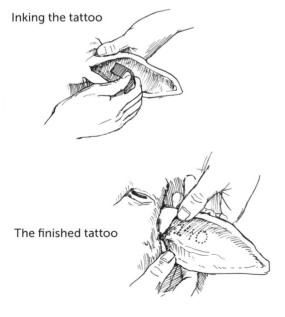

Inking the tattoo

The finished tattoo

When ear tagging or tattooing, it never hurts to give the animal a shot of tetanus antitoxin. After you've done that, remove and disinfect the digits before moving on to another sheep. When you've finished, disinfect the equipment before storing it.

## Tattooing Tips

- Tattoo at least a month before a show or production sale to allow tattoos to heal and be fully legible.

- Tattoo needles must make holes that are large enough to accept an adequate amount of ink or the tattoo will be difficult to read. Brand-new, extra-sharp digits don't do that; lightly filing the tips before using new digits often helps.

- Use small-digit pliers when tattooing lambs as the tattoo will get bigger as the ear grows.

- If a tattoo fades, you'll have to re-tattoo that ear. Check with your registry regarding their policies on reapplication of tattoos.

- When it's difficult to read the tattoo in a dark-pigmented ear, scrub the ear with alcohol to remove grease and dirt, and then shine a flashlight behind the outside of the ear.

## Microchips

Microchips are tiny devices that are implanted under an animal's skin. Based on radio frequency identification (RFID) technology, they contain no power source and are designed not to act until they are acted upon. Microchips are composed of three parts: a silicone chip, a coil inductor, and a capacitor. The silicone chip contains a unique identification number and the circuitry needed to relay that information to a handheld scanner; these components are contained in a biocompatible glass housing about the size of a large grain of rice. The chip is inserted using a syringe with a 12-gauge needle. The process appears to be nearly painless.

To read an implanted chip, the operator passes a microchip scanner over the implantation site. The scanner emits a low radio frequency that provides the power needed to transmit the microchip's code and positively identify the animal. Unfortunately, until recently chips and scanners weren't standardized and not every scanner can read every chip. The new Euro chips, however, can be read by newer scanners manufactured by all three major chip makers, AVID, Destron, and HomeAgain, making them a best choice for compatibility.

Earlier chips had a disquieting habit of migrating away from the injection site, but newer chips incorporate Bio-Bond materials that are fully compatible with animal tissue, making migration problems obsolete.

Chips have a 25-year life expectancy and cannot be altered or easily removed, making them a first-rate permanent means of positive identification. The downsides are that microchipping is comparatively expensive (about $20 to $30 per animal plus veterinary fees if you choose not to implant the chip yourself), and chips can't be read without a scanner.

### IMPLANTING MICROCHIPS

If you are adept at giving injections you can implant your own microchips, or a veterinarian can implant them for you. Each microchip comes in sterile packaging complete with a single-use, pre-loaded syringe and a pack-

Microchip applicator and chip

Plastic link collar

age of stickers printed with the unique code programmed into the chip. These stickers can be affixed to registration papers and similar records, and the code recorded with appropriate registries and databases like the ones managed by AVID and HomeAgain.

Injection sites and methods vary from species to species; sheep are implanted subcutaneously in the manner of implanting dogs and cats, near the base of the ear or between the shoulder blades.

### Temporary ID

Colored plastic neck chains or neck straps can be used to identify sheep according to age, ancestry, birth dates, productivity, and so on, and you can use them to lead tame sheep. To reduce the risk of collared animals hanging themselves on fencing, browse, or another sheep's horns, choose lightweight plastic models that easily break in an emergency.

For very temporary markings use chalk or wax markers, paint brands (inexpensive numerical branding irons dipped in marking fluid), paint sticks, or aerosol spray paints designed for use on livestock. However, some of these products are *not* scourable, meaning they can't be removed from shorn wool; read the label before you buy. It's important.

Chalk or paint sticks, wax markers, and aerosol spray paints last about one, four, and up to six weeks, respectively. They're easy to read from a distance and are great for marking animals in a flock as you vaccinate or worm them. Some producers use spray paint to mark ewes and their lambs with matching numbers. The first family is marked with a 1, the second with a 2, and so on; this has the added advantage of giving the producer the approximate age of each lamb at a glance.

# Henry Stewart's Hints for New Shepherds

Beginners, take note! These wonderful tips from a sheep book published in 1900 are as useful today as they were 100 years ago. You can't go wrong following this advice.

- Select that breed of sheep to which you take a fancy, for what one admires or loves the most, he will give his mind to the most.

- It will be safest to begin with a few of the common sheep for the first year, and then rear the lambs, and the next year get such a ram as you think you would like the best.

- Get a ram two years old next spring. Get a good one. Never mind a few dollars in the cost of it. Ten good lambs will pay all the cost of as good a ram as you wish to have.

- You won't go wrong on any breed, if it is a good animal, well bred and healthy.

- Get no ewes under three years of age. Young ewes need better care than a beginner can give.

- Make friends with your sheep. 'The good shepherd loves his sheep and they will follow him.' But they won't follow any one who ill uses them.

- Don't put them in confinement, but give them an open shed in which they may go as they wish, in or out. They will know to go in when it rains, which is more than some people do.

- Dry litter and plenty of it, will keep the floor from smelling although the manure may lie in it all the winter. And this is advisable.

- Feed regularly, at the same hour every day. . . . Give the sheep all the salt they will eat.

- Take good care of the ewes that are carrying lambs. Don't let them get crowded, or chased, or punched by cows, and don't let them get moldy stuff to eat.

- Preserve the fleece from dust or litter from the floor overhead. Keep the wool as clean as possible. Remove the locks of wool from behind the sheep, and this especially applies to ewes when the lambs come.

- Be patient, kind, watchful, attentive, prompt, thoughtful, and above all other things, be regular to the hour in feeding and watering. Sheep don't, as a rule, carry watches, yet they are watchful, and know the time of day, and if they are not attended to they will let you know by their bleating. Don't wait for this, set your times; the sheep will soon know them; and be particular to be on time every time. A fretful sheep will soon be a sick one, and a sick one is apt to be a dead one in a short time. Keep your sheep happy and they will make you happy.

— Henry Stewart, *The Domestic Sheep; Its Culture and General Management* (1900)

A SHOWCASE
OF SHEEP

# A BREED SAMPLER

Balwen Welsh Mountain

Barbados Black Belly (Caribbean)

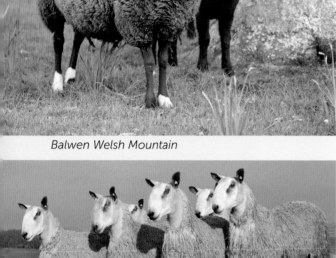

Blue-faced Leicester (United Kingdom)

Navajo-Churo (United States)

Hampshire Down (United Kingdom)

Icelandic

Jacob (United Kingdom)

Tunis (United States)

Kerry Hill (Wales)

Scottish Blackface

Swaledale (United Kingdom)

Wensleydale (United Kingdom)

# HORN SHAPES

Castlemilk Moorit (Scotland)

Herdwick (United Kingdom)

Jacob (United Kingdom)

Manx Loaghtan (Isle of Man)

Swaledale

Racka (Hungary)

Scottish Blackface

# FLEECES UP CLOSE

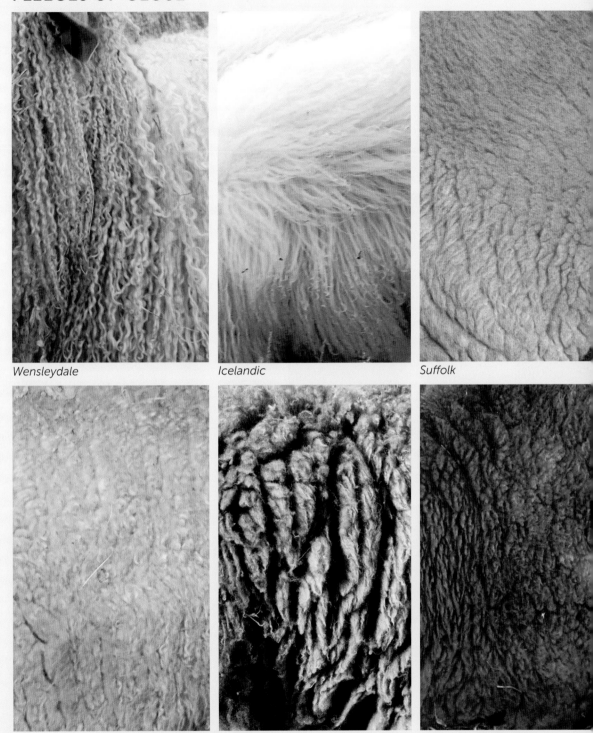

Wensleydale

Icelandic

Suffolk

Classic/Miniature Cheviots, showing variations in wool

# FOR THE LOVE OF LAMBS

# SHEPHERDS AROUND THE WORLD

**LEFT TO RIGHT**

**Top row:** *New Zealand, Kyrgyz Republic, Iceland, Australia*

**Middle row:** *Algeria, India, Nepal*

**Bottom row:** *Mongolia, Afghanistan*

# ANIMAL GUARDIANS

This page, top: *Llama (Canada); Polish Lowland Sheepdog*
bottom: *Akbash dog (Montana)*

Opposite page, top: *Llamas (North Carolina)*
middle: *Maremma dog (New York); Donkey (Portugal)*
bottom: *Great Pyrenees (France)*

# SEPARATING THE SHEEP FROM THE GOATS

Many people have trouble telling sheep from goats. Don't laugh! Some hair sheep breeds are very goat-like and Angora goats resemble woolly sheep. Their voices differ, however:

Sheep generally stick to some variation of "baa," while goat calls vary by breed from sedate "mehs" to ear-splitting screams. Here are some other similarities and differences between them.

**LIPS**

The ovine upper lip is divided by a distinct groove, while the hircine lip has only a superficial groove.

**SCENT GLANDS**

Sheep have scent glands in the lower corners of their eyes and between their toes.

Goats have scent glands behind and slightly to the sides of their horns.

Merino Ram

Angora Goat

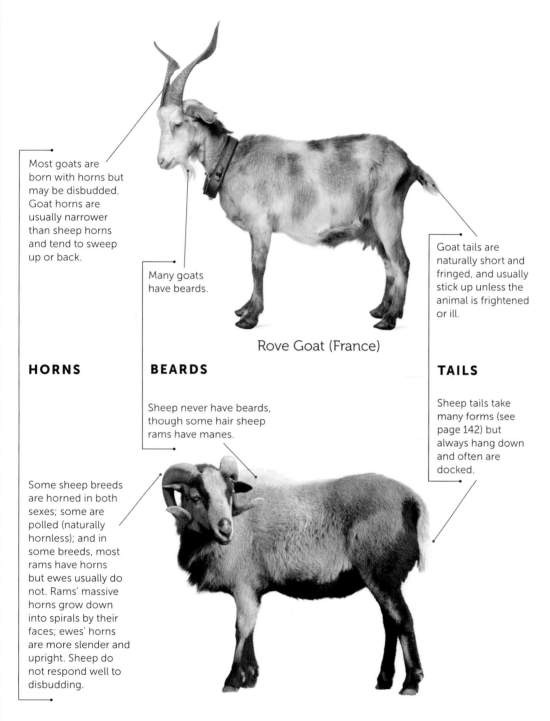

Most goats are born with horns but may be disbudded. Goat horns are usually narrower than sheep horns and tend to sweep up or back.

Many goats have beards.

Goat tails are naturally short and fringed, and usually stick up unless the animal is frightened or ill.

Rove Goat (France)

## HORNS

## BEARDS

## TAILS

Sheep never have beards, though some hair sheep rams have manes.

Sheep tails take many forms (see page 142) but always hang down and often are docked.

Some sheep breeds are horned in both sexes; some are polled (naturally hornless); and in some breeds, most rams have horns but ewes usually do not. Rams' massive horns grow down into spirals by their faces; ewes' horns are more slender and upright. Sheep do not respond well to disbudding.

Barbados Black Belly Sheep

# SHEARING

An experienced shearer can strip the wool off a sheep in 3 to 5 minutes with electric clippers.

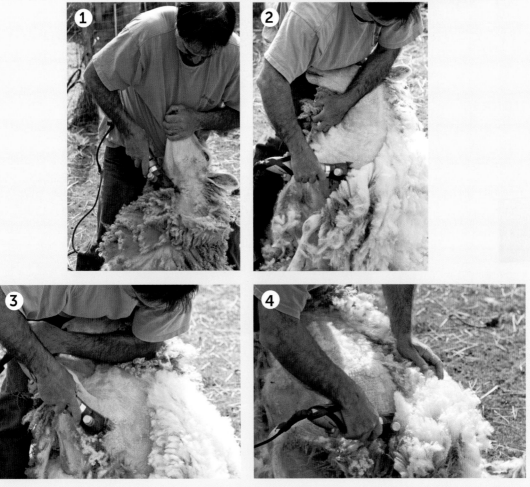

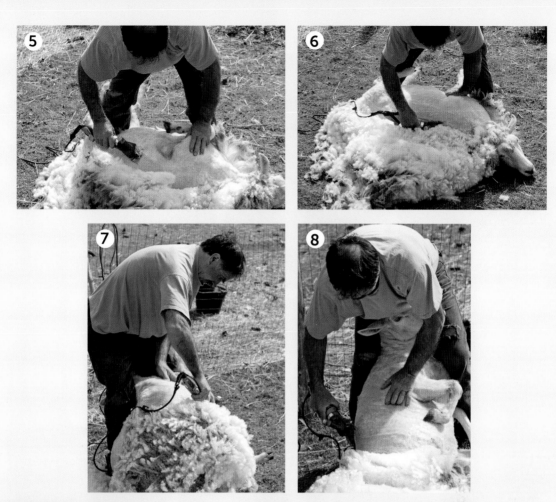

# MEET A FEW OF MY SHEEP

Jacy

Athabaasca

Angel

Rumbler

Hope

Miss Maple

Baamadeus

Louie

Shebaa & Fosco

Othello

# Part 2

# Keeping a Few Sheep

CARRO DE CARNEIROS PARANÁ

F 64

Pet Lamb
and Sod
House.

*This postcard of a girl and her pet lamb was mailed on
September 2, 1912, and is addressed to Mrs. Cliff Wolcott
of Danville, Illinois.*

# CHAPTER 5

# Housing Sheep

*Last night I counted five thousand sheep in
those three beds. So I had to have another bed
to sleep in. You wouldn't want me to sleep
with the sheep would you?*

— Groucho Marx

You don't need a lot of stuff to keep sheep. There are, however, a few essentials such as housing, pens, good fences, feeders, water containers, halters, and leads. You have to feed them too, of course, but we'll talk about that in chapter 6.

## Providing Housing

Basic sheep housing is the essence of simplicity. Sheep need a draft-free, dry place to get out of the sun and weather; that's it. Technically, adult sheep get by nicely in tree-shaded pastures except in the very far North; no artificial structures needed. However, shelter from storms and shade during hot summer days are readily utilized by pastured sheep. And newborn lambs and newly shorn sheep do need protection from wet, chilly weather, even in balmy climates.

Sheep can, of course, be housed in traditional barns and pole buildings, but a few sheep can also thrive in a small, three-sided loafing shed, a refurbished chicken coop, a homemade A-frame, a hoop house made of cattle panels and tarps, a calf hutch, or a box stall in the horse barn. Some of our sheep and goats live in homemade loafing shelters and the rest have Quonset-style Port-a-Huts. If we could start over, we'd buy enough Port-a-Huts for all of them. They're inexpensive yet tremendously sturdy, and cleanup is as easy as picking up the hut and hauling it to a new location. Whatever you choose, make sure you can get inside it or take it apart to clean it.

An adult sheep of a full-size breed needs at least 10 square feet (0.93 sq m) of living space under shelter and free access to pasture or 20 square feet (1.86 sq m) of outdoor exercise

A commercial Port-a-Hut

area. Sheep with full-time access to pasture or a large exercise area can crowd closer together indoors during inclement weather, but not for long. Overcrowding leads to heavy parasite loads and stress-related disease.

## Location and Ventilation

Whatever type of shelter you choose, it must provide a draft-free sleeping area and proper ventilation. Housing must be situated on an easily accessible, well-drained site. Field shelters and Port-a-Huts should face south away from the prevailing wind. Electricity and running water are huge plusses.

Sheep kept in drafty structures or poorly ventilated, tightly enclosed winter housing are prone to respiratory ailments such as pneumonia and bronchitis. Take a deep breath; if you smell ammonia, ventilation is inadequate.

Wool sheep can handle subzero temperatures as long as they're dry. If the weather turns nasty right after shearing or if an old or sick sheep gets cold, buy (or make) him a sheep coat rather than shutting all the windows and doors.

**LEARN FROM THE PAST**

The shelters for sheep are variously constructed to suit the taste or circumstances of the flock-master. In all cases, however, thorough ventilation should be provided, for of the two evils of exposure to cold or too great privation of air, the former is to be preferred. Sheep cannot long endure close confinement without injury. In all weather, a shed closely boarded on three sides, with a tight roof, is sufficient protection; especially if the open side is shielded from bleak winds, or leads into a well-enclosed yard.

— R. L. Allen, *Domestic Animals: History and Description of the Horse, Mule, Cattle, Sheep, Swine, Poultry, and Farm Dogs* (1848)

Floors can be built of concrete, stone, packed dirt, sand, or clay, and bedded with hay, long-stem straw, or covered with rubber stall mats (avoid sawdust, shavings, and chopped straw for bedding wool breeds; these materials end up in fleeces). Some people build sleeping platforms for their sheep. Our platforms are made of heavy exterior plywood fastened to free wooden pallets from our feed store. We use them for flooring in our loafing sheds and Port-a-Huts and cover them with straw.

### Storage Areas

You'll need weather-tight, indoor space to store feed and bedding along with the everyday paraphernalia you need to keep sheep. The room or building should be fitted with a secure closure and carefully closed every time you leave the area, so sheep can't get in, eat their fill, and die of deadly bloat or enterotoxemia (unfortunately, this happens far more often than you probably think).

---

### Warm Winter Sheep Coats
. . . . . . . . . . . . . . . . .

Goat supply companies such as Hoegger Supply Company sell horse-style winter blankets tailored for goats that fit sheep every bit as well.

Or visit a Goodwill store or shop garage sales to buy secondhand sweatshirts and cardigan sweaters sized to fit sheep (wool cardigans are the *crème de la crème* of ovine couture). Trim off the sleeves, feed the sheep's forelegs through the armholes, and button the sweater up his back. Done!

---

Feed and bedding must be kept dry. It also needs protection from hungry rodents and farm pets such as cats or dogs that might sleep in or soil it.

Place feed, especially grain, in tubs with securely locking lids. Or even better, haul in a nonfunctional, chest-style freezer for grain and mineral storage. Find one through your local Freecycle chapter or ask at appliance dealers; you may be able to get one for free. (Just be sure to disable the lock so that a child who happens to be playing hide-and-seek doesn't get stuck inside.)

## Constructing Pens

If you utilize a larger structure like a barn or a garage, you'll need pens inside to contain your sheep. Enclose pens with solid partitions anywhere drafts are an issue. Alternatively, use welded-wire cattle or sheep panels for great ventilation during the summer months, and cover the panels with plastic tarps when winter arrives.

I prefer cattle panels for building pens, exercise runs, and small- to medium-size outdoor paddocks. Cattle panels, sometimes called *stock panels*, are prefabricated lengths of mesh fence welded using galvanized ¼-inch (6 mm) rods. Most cattle panels are 52 inches (132 cm) tall, a good height for keeping predators at bay. Horizontal wires are set closer together near the bottom of the panel to prevent small lambs from wriggling through. They're sold in 16-foot (approximately 5 m) lengths that can be trimmed to size using heavy-duty bolt cutters.

Sheep panels are cattle panels manufactured in 34- and 40-inch (86 and 102 cm) heights.

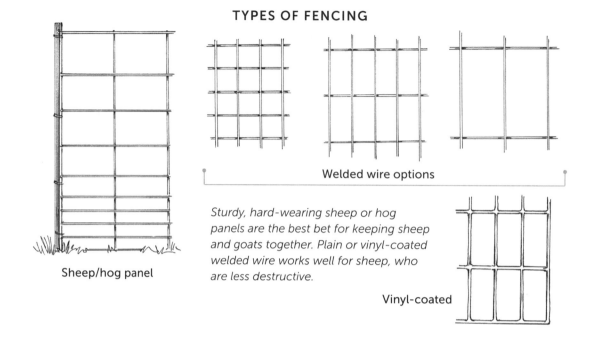

Welded wire options

Sheep/hog panel

*Sturdy, hard-wearing sheep or hog panels are the best bet for keeping sheep and goats together. Plain or vinyl-coated welded wire works well for sheep, who are less destructive.*

Vinyl-coated

Their horizontal wires are set more closely together than they are in cattle panels. They're ideal for interior dividers but not tall enough to provide predator protection as exterior fencing, though they work well when they're beefed up with several strands of supplementary electric fencing above the panels.

The downside of cattle and sheep panel construction is that the raw end of each rod is very sharp. To render these panels more user friendly, file the end of each rod to smooth its razor edge.

Premier1 sells panels designed for sheep applications. They come in 36- and 40-inch heights (91 and 102 cm) and in 4- to 6-foot (1.2 to 1.8 m) lengths that are just right for building V-type hay feeders. Premier1's panels or cut-to-size standard cattle panels are also useful for building walk-through gates, lambing jugs, and small-size temporary pens. As a bonus, the rod ends on Premier1 panels are pre-smoothed at the factory.

## Building Fences

Probably the hardest part of shepherding is protecting your sheep from predators; it calls for extra-sturdy fences. Two main types of fencing should be considered:

*Perimeter fences* are strong, permanent fences erected around a property or grazing area. They are the first line of defense against predators. *Cross fences* (or interior fences) are used to subdivide large areas into smaller paddocks. They can be built as permanent or temporary installations.

### Cattle Panels

Several types of farm fences work well with sheep, but I especially like cattle-panel fences.

They're strong, durable, and easy to erect: set wooden or steel fence posts, carry the rigid panels into place, wire them to steel posts or tack them to the inside of wooden ones, and voilà, instant fence. Unfortunately, enclosing large areas with panels is costly. A close second-best is quality woven wire.

## Woven Wire

Woven wire, also called *wire mesh* or *field fence*, is made of horizontal strands of smooth wire held apart by vertical wires called *stays*. Horizontal wires are usually set closer together near the bottom of the fence. High-tensile woven wire costs more than standard woven wire fencing, but it's rust-resistant, sags less, and is lighter in weight. Correctly installed woven wire is the most secure type of affordable small livestock fencing; it's the one we use as perimeter fencing on our farm.

A 48-inch (122 cm) woven wire fence with one or two strands of electric wire along the top, set on fence posts placed 14 to 16 feet (4 to 5 m) apart, makes a stellar perimeter fence for sheep.

When buying woven wire, check the tag; the numbers printed on it will tell you its characteristics. For instance, 10-47-6-9 fencing has 10 horizontal wires, is 47 inches (119 cm) tall, has a 6-inch (15 cm) spacing between stay wires, and is made of 9-gauge wire.

## Barbed Wire

Properly installed barbed wire holds sheep but is not as predator-proof as cattle panels or woven wire, and for safety reasons it must never be electrified to give it more bite. It takes eight to ten strands of strong barbed wire fence to contain sheep and keep predators at bay.

### CHOOSING FENCE POSTS

Wooden posts come in treated and untreated varieties. Treated posts will last 20 to 30 years; untreated ones from 2 years up, depending on the type of tree they're made from.

Steel posts, commonly referred to as T-posts, are fireproof, long-wearing, lighter in weight than wood, and relatively easy to drive. They also ground the fence against lightning when the earth is wet. They tend to bend if larger livestock leans against them or you back over one with your truck. Unbent T-posts should last 25 to 30 years.

Tape- and net-style temporary fencing is often erected using step-in fiberglass posts. Choose fiberglass posts with built-in UV protection; unprotected posts degrade as they age, leaving fiberglass splinters under your skin if you're not careful when handling them.

## Electric Fencing

Because wool effectively lessens the jolt, electric fencing isn't the best type of fence for containing wool sheep and shedding breeds. It does work well for hair sheep, though. It comes in aluminum, regular steel, and high-tensile steel. For permanent fencing, high-tensile lasts longest, but aluminum wire works well, too. Aluminum wire is rustproof, it's easily shaped with bare hands, and it conducts electricity better than plain steel wire.

Five to seven strands of electric wire is adequate to contain most sheep, though not necessarily to keep predators at bay. It's best used for interior rather than perimeter fencing.

# Preventing Predation

Predation is a serious problem and one you'll have to address in order to keep sheep anywhere in North America. Sheep are killed mainly by coyotes but also by dogs (usually free-roaming pets rather than feral packs), mountain lions, bears, foxes, eagles, bobcats, and wolves. The concept of livestock guardian animals goes back about 6,000 years to Turkey, Iraq, and Syria, where dogs were first trained to protect sheep and goats. Nowadays savvy livestock owners still use guardian dogs to protect their livestock, but they've also added guardian donkeys and llamas to the mix.

**USING A GUARDIAN DONKEY.** Most donkeys have an innate hatred for anything that (to them) resembles a wolf; they require no specialized training and will give chase while braying, biting, and sometimes pawing or kicking at canine invaders. They are hardy animals and typically easy keepers. You can usually buy a guardian-quality standard-size or larger donkey for $100 to $800. Not all donkeys are suitable guardians, and intact males never are. It's best to choose a donkey with prior sheep or goat guarding experience.

Miniature donkeys are adorable, but if you buy one, you'll just be adding to the predator buffet.

**USING A GUARDIAN LLAMA.** Llamas require the same food, vaccinations, and hoof care that sheep do. Many make excellent guardians; others don't. Never use an intact male llama to guard sheep; he could

seriously injure or even kill a ewe by trying to breed her. A guardian-quality, gelded llama costs from $200 to $750; females usually sell for somewhat more.

Llamas are in their element when guarding livestock against less aggressive predators such as foxes or a dog or two. They are not effective at dealing with "big guns" such as mountain lions or bears or with packs of aggressive canines, all of which can easily kill or maim a guardian llama.

**USING A LIVESTOCK GUARDIAN DOG (LGD).** These dogs have been bred for generations to live with and identify with their charges rather than hunt them. Wolf-like characteristics such as pointy ears and muzzles and dark coloring have been eliminated, which makes sheep more accepting of them. All of the livestock guardian breeds are large, intelligent, strong-willed, and potentially aggressive and dominant dogs, so they aren't to be casually handled by the faint of heart. Before buying a livestock guardian dog, visit LGD breeders and livestock owners who keep them. Compare philosophies and training methods. Know what you're getting into before you bring a dog home.

Most are aggressive toward other dogs; they sometimes kill household pets who wander in with their charges, and some will fight to the death with other guardian dogs of the same sex. Since most predators strike at night, livestock guardian dogs are most active after dark. Barking is their primary means of warning off intruders, so expect a lot of nighttime barking from your LGD.

To learn more, read *Livestock Guardians: Using Dogs, Donkeys, and Llamas to Protect Your Herd* by Janet Vorwald Dohner.

Electric fences are primarily psychological rather than physical barriers, so you must train sheep to respect them. Place untrained sheep in a fairly small area electrified by a powerful fencer, and then entice them from the sidelines with a pail of grain. Once they've been zapped, most sheep avoid electrified fences like the plague.

## Portable Fencing

Net-style portable electric fencing is economical, easy to use, and wildly popular with grass-based farmers, people who use sheep for brush control, and shepherds who practice controlled grazing. It can be charged by battery or solar panels. Some styles come with built-in fence posts and most roll up onto easy-to-use reels; however, net fencing has one serious drawback.

Unless horned sheep are trained to respect electric fences before encountering soft net fencing, they sometimes tangle their horns in the mesh; then they're repeatedly shocked, they stress out, and some even die. The same thing happens when lambs push their heads through openings in the fence. Train your sheep to respect electric fences before you use them.

## Building Better Electric Fences

**CHARGE.** Buy an adequate fence charger. The box will tell you how many miles of fence it charges, but take that information with a large grain of salt. The more powerful your fencer, the fewer problems you'll have, so pick a model that really packs a punch.

**GROUND.** According to University of Maryland Extension Sheep and Goat Specialist Susan Schoenian, an estimated 80 percent of the electric fences in the United States are improperly grounded. Follow grounding instructions printed in the user's manual that comes with the charger.

**QUALITY.** Use quality insulators. Choose a brand treated to resist damage done by ultraviolet (UV) light. Don't skimp on wire. The larger the wire, the more electricity it can carry. Buy a voltmeter and check your fences every day. A good one costs $50 to $75.

**SPACE.** Don't space wires too closely, even when enclosing wool sheep. To get the most from your energizer, set them at least 5 to 7 inches (13 to 18 cm) apart.

**CAUTION.** Never electrify barbed wire.

Electronet fencing

Here are some good uses for portable net fencing:

- To subdivide pastures for rotational grazing
- To erect small pastures near the barn for special-needs animals such as rams or elderly sheep and for limited grazing for animals in quarantine
- To keep coyotes and marauding dogs out (a few brands also work well for foxes)
- To keep livestock guardian dogs in
- To fence steep, rocky, or otherwise uneven land
- To use as boundary fences on rented land
- To make lanes for moving sheep without assistance

## Everyday Equipment

In addition to proper shelter and sturdy fencing, you'll need everyday items such as halters, leads, feeders, watering devices, and a way to haul your sheep. Here's what you need to know.

### Halters, Collars, and Leads

You'll need a means of leading and restraining your sheep. We prefer the type of web halters used on alpacas and llamas, not the flat nylon halters sold as "sheep halters" that are designed for leading but not tying, or the adjustable rope halters that tend to pull up into the eyes of small-breed sheep.

The best halters we've used are made of sturdy nylon web, very adjustable, and crafted using solid brass fittings. They're designed by and are available from llama and alpaca clinicians Cathy Spaulding and Marty McGee

Bennett (see Resources). Be sure the halter you buy fits your sheep's head. It shouldn't squash his face but too loose is wrong, too; strongly Roman-nosed breeds are especially difficult to fit. Use this type of halter only for leading or restraining sheep; it should never be left on full-time.

Another option, particularly for hair sheep or newly shorn wool sheep, is the type of plastic link or lightweight leather or nylon collars used on goats (see picture, page 47). Unlike halters, these are usually left on the animal at all times. Don't, however, use strong, double-ply collars for everyday use; if a collar catches something and hangs up your sheep, you want it to break. We use cheap, single-ply nylon dog collars from the local dollar store and plastic chain goat collars from Hoegger Supply Company. The chain collars fasten with plastic connectors that pull apart, releasing the animal if the collar is snagged and its wearer pulls back. If you can find the collar, usually you can put it back on the sheep.

## Feeders

All good sheep-feeding apparatus is designed to discourage waste. Sheep, being fastidious creatures, won't touch hay, grain, or minerals they or any other animal has peed or pooped in. However, they don't think twice about peeing and pooping in their own feed.

## HALTERS AND COLLARS

*Sheep halters come in several different styles. Some are more suitable for leading, others for tying.*

A hay rack/tub feeder

Don't feed hay or grain directly on the ground; it contributes to parasitism and disease, not to mention wasted feed.

Consider using grain feeders you can hang on the fence and remove after your sheep have eaten; if feeders aren't there, they won't get pooped in. Another approach is to mount grain feeders 6 inches (15 cm) higher than your tallest sheep's tail and provide booster blocks or rails for your sheep's front feet to stand on.

If you aren't concerned about fiber quality (we'll talk about that in chapter 6), invest in a combination hay rack and tub-type feeder so that "wasted" hay falls in the tub instead of on the ground, giving diners a second chance to eat it. Allow 16 to 18 inches (41 to 46 cm) of feeder space per average horned sheep and 12 inches (30 cm) if your sheep are polled. Feeders should be large enough, or you should have enough of them, so all of your sheep can eat at the same time; otherwise timid individuals may not get their share.

Or make your own inexpensive fence line hay feeder by wiring the bottom of 4- by 4-inch (10 x 10 cm) mesh welded-wire panels to an existing fence and adding sturdy wire spreader arms at the top.

### HAY BAGS FOR SHEEP

Need an inexpensive way to feed hay to a few head of sheep? Use a hay bag designed for feeding horses. Not a hay net, but rather a *bag* made of solid fabric with an opening in the side to access hay. Horned sheep snag their horns in knotted cord hay nets and sheep of all kinds can catch their legs in them, so don't use the nets.

## Watering Devices

A sheep drinks ¼ to 4 gallons of water per day depending on her size, physical condition, level of activity, what sort of feed she's eating (juicy spring pasture provides much of an animal's fluid intake), quality of the water, temperature of the water, and temperature of the environment. Lactating ewes have the highest requirements, and to prevent urinary calculi, rams and wethers must drink a lot of water, too.

When water is contaminated with droppings, algae, dead birds or bugs, leaves, and other debris, however, sheep drink only enough fluid to survive. Dump questionable water on a daily basis and scrub out buckets and tubs before refilling. Choose a group of small containers over one big tank; smaller ones are easier to clean, and if one container becomes contaminated, other sources of water will still be available.

During the summer months, place water tubs and buckets in the shade. This helps inhibit algae growth, and because the water stays cooler and fresher, the sheep drink a lot more. When it's *really* hot, freeze ice in plastic milk bottles and submerge one in each trough or tub around noon; they'll cool water for several hours during the heat of the day and your sheep will appreciate this treat. Refreeze them overnight and they'll be ready to use again the next day.

### WATERING TUBS FOR CHEAP

If you have cattle, you probably have empty plastic mineral lick tubs sitting around. Smaller ones make first-rate grain or mineral feeders and dandy lamb watering troughs. Larger ones are handy for watering adult sheep. If you don't have tubs of your own, post a request to your local Freecycle chapter (see Resources); generous farmers and ranchers will give them to you for free.

Sheep cannot ingest enough fluid by eating snow. During the winter, keep water supplies from freezing with bucket or stock tank heaters. Be sure to sheath the cords in PVC pipe or garden hose split down the side and taped back together with duct tape; animals have been known to gnaw through electrical cords and die.

Where lambs are or will be present, shallow containers are the rule. Ewes sometimes give birth standing up and can drop their newborns into a pail of water where they drown. Lambs might leap or tumble into buckets or tubs and drown. *Never* use five-gallon recycled plastic food service buckets or any other narrow, deep water containers in lambing areas or jugs. Several shallow water containers are better than one deep one.

### Ewe Haul

Sheep can be safely hauled in horse trailers, goat totes (a type of cage that slides into the bed of a pickup truck), truck toppers, minivans, SUVs, and just about any vehicle that is big enough, is properly ventilated, provides safe footing, and is escape proof.

Escape proof is important. Stressed sheep can leap higher and squeeze through smaller openings than you can imagine. Err on the side of caution. If a sheep bails out while you're driving down the road, she's sure to be killed.

Restrain sheep when hauling them in vans and SUVs; you don't want one jumping into the driver's lap. Extra-large dog crates are perfect for hauling miniature adults and lambs of all breeds. Secure larger sheep with halters and leads; they travel best facing the rear of the vehicle or sideways, that is, the sheep facing the side of the vehicle.

# Feeding Sheep

*The sight of a pasture it loves stirs in a sheep the desire to feed: show it a stone or a bit of bread and it remains unmoved.*

— Epictetus

I can't tell you precisely what to feed your sheep. Animals' nutritional needs depend on a lot of variables including their age, sex, level of productivity (lactating ewe, idle pet, hardworking ram), and the types of feed available where you live.

I do, however, recommend a diet based on forage (pasture or hay) augmented with a sheep-specific mineral supplement formulated for the type of forage you feed and the area where you live. Fine-tuning a flock's dietary needs is a daunting task beyond the scope of this book, so I strongly suggest meeting with your County Extension agent to discuss a diet based on your flock's needs and the locally available feeds.

Another option: read the feeding section on the Sheep 201 website (see Resources) by University of Maryland Extension Sheep and Goat Specialist Susan Schoenian. Then you can do your own math.

Whatever diet you choose, keep in mind that the ovine digestive tract was designed to break down the cellulose in forage, and it doesn't adapt particularly well to starchy, high-grain diets. Grain-rich diets predispose sheep to serious nutritional diseases such as acidosis, bloat, enterotoxemia, laminitis, listeriosis, polioencephalomalacia (goat polio; sheep can get it, too), and urinary calculi. In fact, some classes of sheep, such as dry ewes and wethers, especially of the hardier and more primitive breeds, thrive on a diet of forage alone. For these sheep, mineral supplements can be used to balance the ration without exposing them to the dangers of grain overconsumption.

## Your Sheep's Digestive System

All true ruminants, such as sheep, goats, cows, deer, and bison, have stomachs with four chambers: the rumen, reticulum, omasum, and abomasum.

## NECESSARY NUTRIENTS

| Component | Provides | Sources | Increase dose during |
|---|---|---|---|
| Carbohydrates | Energy<br><br>Maintenance of body heat, exercise, growth, reproduction, lactation | Green pasture, high-quality hay, grain (in moderation) | Lactation, growth, drought |
| Protein | Growth, body maintenance, reproduction (especially last 4-6 weeks of gestation), lactation | Green pasture, high-quality hay (especially legumes like alfalfa)<br><br>Whole soybeans, soybean meal, black oil sunflower seeds, sunflower meal, whole cotton seeds, cottonseed meal, peanut meal, canola meal, alfalfa pellets<br><br>Commercial protein supplements, protein blocks, high-protein bagged feed (all in strict moderation) | Late gestation, lactation, growth |
| Minerals | Bone growth, milk yield, general growth of lambs, body functions | Commercial mixtures formulated for sheep, salt | Salt and mineral supplements should be fed free-choice at all times |
| Vitamins | Support the entire process | Green pastures, high-quality hay | Growth, reproduction |
| Water | Essential for health productivity, lactation, body functions | Fresh, clean water (containing no droppings, bugs, thick algae, or other muck), up to 3 gallons per sheep per day | Lactating ewes, rams and wethers, unusually hot and steamy weather |

# A SHEEP'S DIGESTIVE SYSTEM

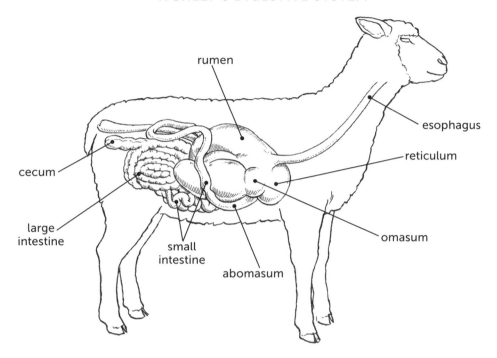

The first and largest of these chambers is the **rumen**. Located on the left side of the sheep's abdomen, it acts as a fermentation vat. A rumen contains billions of bacteria, protozoa, fungi, molds, yeasts, and other tiny microbes that feed on carbohydrates. Microbes convert carbohydrates into volatile fatty acids, which represent the sheep's primary source of energy.

As feed mixed with saliva enters the rumen, it separates into layers of solid and liquid material. Later, when the sheep is resting, he burps up some gas (that's why sheep's breath smells funky) and a bolus of food (cud) and re-chews it more slowly. Then he swallows the food again, repeating the process several more times.

Chewed-up feed flows back and forth between the rumen and the next chamber, the **reticulum**, by way of an overflow flap. Particles remain in this area for 20 to 48 hours; fiber fermentation is a slow process. Eventually the material passes from the reticulum through a short tunnel into the **omasum**.

The omasum is divided by long folds of tissue somewhat like pages in a book. It's lined with tiny, fingerlike projections called *papillae*, which increase its working surface. The omasum decreases the size of feed particles, removes excess fluid from digestive slurry, and absorbs volatile fatty acids that weren't absorbed in the rumen.

The fourth chamber, the **abomasum**, is considered the sheep's true stomach because it's the most like a human stomach. The walls of the abomasum secrete digestive enzymes and hydrochloric acid. Protein is partially broken down in the abomasum before material moves on to the small intestine.

As semi-digested feed flows into the small intestine, secretions from the liver and pancreas are mixed in, making it possible for the enzymes in the small intestine to reduce remaining proteins into amino acids, starch to glucose, and complex fats into fatty acids. As this occurs, muscular contractions continually push small amounts of material through the system and into the large intestine where bacteria finish digesting the mix.

## Feeding Hay

Whether or not you grain your sheep, you should feed them good hay when they aren't on pasture. Given the choice between small square bales (they're actually rectangular, not square) or big round ones, use small square bales; they work best for sheep. Small squares separate naturally into flakes, so you can feed exactly the amount sheep need. Large round bales are okay if they've been stored under cover and are free of mold, but your sheep will pull down more hay than they eat, then poop and lie in it. You can also expect a lot of debris-contaminated fleece at shearing time.

Nutrient-rich hay is green. Quality alfalfa is dark green. Good grass hay is light to medium green. The outsides of both legume (alfalfa, red clover, white clover, vetch, birdsfoot trefoil, peanut, soybean, lespedeza, cow pea) and grass (Bermuda grass, bromegrass, timothy, orchard-grass, bahiagrass, ryegrass, fescue, bluestem, Reed canarygrass, Kentucky bluegrass, native grasses) hays become bleached if exposed to light while they're in storage, but inside of the bales the hay should still be green. Hay that is pale to golden yellow both inside and out was sun bleached in the field. Nutrition-poor and apt to be dusty, such hay is not a good buy.

### Tips for Buying Hay

Quality hay is composed of leafy plants mown just before they're due to bloom; or, depending on type, sometimes while in early bloom. It has thin, flexible stems that easily bend without breaking. Coarse, stemmy hay is low in nutrients, and sheep don't like it so they waste a lot. Most of a plant's protein is in its leaves, while stems are mainly low-energy cellulose.

Never buy hay containing sticks, rocks, dried leaves, weeds, and insects or animal parts. The weeds might be toxic and their seeds will certainly infest your property. Desiccated animal parts, even small ones, can cause deadly botulism, so if you find an animal part in your hay, discard the whole bale.

It's best to buy nutrient-tested hay from a reputable hay dealer. Be extra cautious if you buy hay at auction. Some sellers pile their best bales on the outside of the stack and junky hay deep inside, out of sight.

Inspect the hay you plan to buy. Pay for two or three bales you can open. See what they look like inside, then if the hay doesn't pass muster you aren't obligated to buy the rest.

Buy only the amount of hay you can store under cover. Stack it on pallets, old tires, or poles to get it off the floor; if you don't, the bottom layer will rot. Don't let barn cats use your haystack for a litter box or have kittens in it; cats are primary carriers of toxoplasmosis. Finally, to minimize waste and prevent disease, never feed hay on the ground.

## Give the Sheep a Beer

During spring 2011, our huge Boer goat wether, Salem, apparently ate some bad stuff from the outside of a big bale of hay. (Lesson: discard bad hay where livestock can't reach it.) He developed a bellyache, stopped eating, and became depressed.

After he was on his feet again, we gave Salem all the things said to stimulate appetite and support the gut: vitamin B complex shots, yogurt, and Nutra-Drench. They helped, but not a lot.

Then I posted about Salem to my Hobby Farms Sheep list. Alice, a British member, suggested giving him dark lager. A 90-year-old Welsh shepherdess had recommended beer for sheep with upset tummies, and Alice used it with great success.

Salem, however, wasn't impressed. He spit and sputtered as we carefully poured it down his throat directly from the bottle (beer foams too much to use a dose syringe). But that evening he seemed much better and two beers later, he was tucking into feed just fine.

We've since dosed another sick goat (who licked his lips and drank his second and third beers from a bowl) and several sick sheep. And by cross-referencing vintage livestock books I learned that beer is a traditional pick-me-up for animals who appear to feel "off" and don't want to eat.

So, try it when you have a peaked sheep. The traditional dose is a beer every two to four hours. Is it the hops that do the trick? Who knows? But it can't hurt — and we think it works!

*You can make a drench from a bottle of beer by fitting the nipple from a calf feeding bucket over the neck of the bottle (enlarge the opening slightly so the beer will flow). I recommend some additional padding around the neck of the bottle to prevent the sheep from biting down on the glass.*

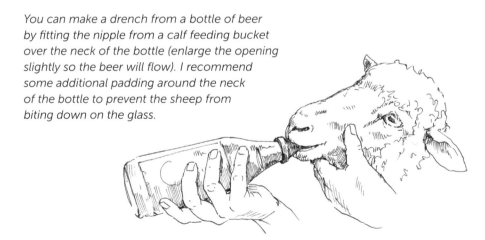

## Tips for Feeding Hay

Don't feed uncured new crop hay. Fresh-from-the-field hay should be aged for 5 to 6 weeks. Uncured hay can trigger bloat.

Musty-smelling hay and hay containing obvious mold or dust was cut and stored before it was fully dry. Even if you don't spot actual mold, dusty hay is full of mold spores and shouldn't be fed.

Old hay isn't necessarily bad hay. Nutritious hay put up dry and carefully stored holds much of its nutrient content for several years. High-quality old hay is a better buy than mediocre new crop hay.

## What You *Must* Know about Copper Toxicity

Dietary copper is an essential and yet potentially toxic mineral required by all farm livestock, including sheep. Sheep require about 5 ppm (parts per million or mg/kg) of copper in their total diet. Unless they live where soils are highly copper deficient, they usually meet those needs through everyday feed. Hay, for instance, contains 2 to nearly 15 ppm of copper (depending on factors including type and stage of harvest), corn has 4 ppm, and oats 6.8 ppm.

However, sheep are also extremely susceptible to copper toxicity. Depending on breed, age, health status, and the levels of other minerals consumed, especially molybdenum and sulfur, toxicity can occur at levels as low as 20 ppm of copper in an individual's diet.

According to Susan Schoenian, University of Maryland Extension Sheep and Goat Specialist, and Mike Neary, Purdue University Extension Sheep Specialist, down-type, medium-wool ewes of British breeds such as Hampshires, Suffolks, and Southdowns are especially vulnerable to copper toxicity.

Copper poisoning can occur after a single exposure (perhaps from eating a large amount of pig feed) or be cumulative over several months' time. Affected sheep become depressed and lethargic; they may grind their teeth, indicating pain; and they're usually very thirsty. Their membranes are pale, turning to yellow as jaundice sets in. Death usually occurs in 1 to 2 days.

Feed, milk replacers, and minerals formulated for poultry, pigs, cattle, and even goats are much too high in copper for sheep to safely consume. Sheep have died from copper poisoning even after grazing fields fertilized with copper-rich pig manure.

This is a topic to discuss with your County Extension agent. In the meantime, never feed your sheep commercial feeds, milk replacers, or minerals that aren't specifically labeled for sheep or that don't say, "no copper added" somewhere on the packaging.

## Feeding Concentrates

Two types of concentrates are available: energy- and protein-rich feeds. Energy feeds that are high in total digestible nutrients (TDN) but low in protein are barley, beet pulp, corn, milo, molasses, oats, rye, and wheat. Protein feeds with at least 15 percent protein are soybeans and soy meal, cottonseed meal, brewer's grains, and alfalfa pellets.

While it's certainly possible to formulate and mix your own grain rations, again, it's best to consult your County Extension agent before doing so. Or buy bagged complete feed formulated for sheep; this is what we do when our late gestation and lactating ewes and our growing lambs need grain.

## Feeding Minerals

Sheep should have access to mineral supplements specifically formulated for sheep. The most important minerals in sheep rations are salt (sodium and chlorine), calcium, and phosphorus. Calcium and phosphorus are necessary for proper growth and development, but high levels of phosphorus in relation to calcium can cause urinary calculi (see chapter 7).

## If You Switch Feeds, Do It Gradually

In conclusion, if the rumen doesn't contain the necessary microorganisms to digest a certain food, sheep can't digest that food. Never make immediate changes in feed type or content, and that includes going from hay to fresh spring grass or browse. Mix the diet, some of the old with some of the new, over 7 to 10 days until their rumens have a chance to adjust. If you don't, good microbes die and toxins proliferate, sometimes leading to deadly enterotoxemia.

## Ewe Said It

### SHEEPY PROVERBS

The bleating sheep loses a bite.

— *Ethiopia*

Many a sheep goes out woolly and comes home shorn.

— *Denmark*

Don't climb a tree to look for sheep.

— *China*

A lazy sheep thinks its wool is heavy.

— *Turkey*

The lamb is a sheep in the long run.

— *Ireland*

A sheepless household starves.

— *Iceland*

[He is] mutton dressed as lamb.

— *England*

Better lose the wool than the sheep.

— *France*

# Doctoring Sheep

*I struck an idea pretty soon. I says to myself,
spos'n he can't fix that leg just in three shakes
of a sheep's tail, as the saying is?*

— Mark Twain, *The Adventures of Huckleberry Finn* (1884)

In this chapter we'll talk about finding a veterinarian and keeping your sheep healthy. We'll also discuss some common sheep ailments and what you can do to prevent and treat them.

## Finding a Sheep-Savvy Veterinarian

Finding a veterinarian for your sheep might be harder than you think. Small animal vets consider sheep, even pet sheep, to be livestock, while many livestock vets don't see enough sheep to know much about them.

Here are some questions to ask when interviewing a prospective veterinarian:

- Are you interested in sheep?
- Do you have experience treating them?
- If not, have you treated goats, alpacas, or llamas?
- Are you willing to learn about sheep?

- Would you object if I researched relevant topics and brought resources to your attention?
- Do you make farm visits or would I have to haul my sheep to your practice?
- What would happen in an emergency? Does your practice respond to calls 24/7?
- Do you allow clients to carry a tab or is full payment due when services are rendered?
- Do you have a payment plan for large, unforeseen bills?

Don't wait for an emergency to choose a veterinarian. Once you find one you think you'll like, schedule a routine farm call or an office visit to get a feel for the way he and his staff handle your sheep. Because you might at some point have to board a sick or injured sheep at the veterinarian's practice, ask to see the facilities. Can you bring familiar feed? Can the patients see but not interact with other animals so they won't become stressed or lonely?

You are the consumer; you will be paying the bill. If a veterinarian turns out to be uninterested in your sheep or seems incompetent, or if he turns out to have an abrasive pen-side manner that bothers you, find another vet.

Once you find a good vet, hang on to him (or her, of course!). Listen carefully, follow treatment protocols, and pay your bill on time. If something he does upsets you, tell him; don't complain behind his back. Good small-ruminant veterinarians are worth cultivating.

## Keeping Sheep Healthy and Well

Sheep try not to show their weakness when they're sick or in pain, so I can't stress how important it is to carefully look them over at least twice a day, even if they're out on pasture. Make certain all sheep are accounted for. Sick sheep often drift away from their flock mates.

If something is wrong, do something about it; don't wait to see if the animal gets better without treatment. If you don't know what's wrong or you're not positive you know how to treat a problem, *get help*.

Police your pastures, pens, and sheep yards on a weekly basis. Check for toxic plants, hornets' nests, broken fences, sharp edges on metal buildings, protruding nails, stray hay bale strings, and anything else sheep could hurt themselves on. Keep mice, rats, opossums, and the like out of feed storage areas. Don't let cats, dogs, or poultry foul your hay.

Keep your sheep in clean, dry surroundings. Provide well-ventilated but draft-free shelter from the elements. Avoid overcrowding; it leads to stress. Furnish an unlimited supply of fresh, clean water. All sheep, but especially rams, wethers, and lactating ewes, need to drink as much water as they can.

Choose sheep-friendly feed; never feed copper-rich products packaged for goats, horses, cattle, poultry, or pigs. Feed quality hay. Discard dusty or moldy bales and bales with dried animal parts in them. If you don't know how to formulate nourishing grain mixtures, feed bagged sheep concentrates. Make feed changes gradually; give your sheep's rumens time to adjust to new types or quantities of feed.

You *must* provide adequate predator protection for your sheep, be it exceptionally good fences, a secure fold where you can shut them up at night, or a livestock guardian dog, donkey, or llama (see Preventing Predation, page 72). Many thousands of sheep are killed by predators, including dogs, every year.

Buy from disease-free flocks and consider having incoming sheep tested for CL and OPP (pages 99 and 102) before adding them to your flock. Check carefully for signs of hoof rot before you buy. Quarantine all incoming sheep (and goats, if you keep both species; sheep and goats share many diseases). House them at least 50 feet (15 m) from other sheep or goats but where the quarantined sheep can see other animals (preferably sheep). Quarantine for a minimum of 30 days, then disinfect the quarantine pen before you use it again.

Discuss vaccination and deworming protocols with your veterinarian. At minimum, vaccinate against enterotoxemia and tetanus. Set up a hospital area for sick sheep; don't house injured or ailing sheep with other sheep. Make every effort to be with ewes when they lamb. Assemble a lambing kit (page 127) and know how to use it.

# SIGNS OF HEALTH AND ILLNESS

| A Healthy Sheep | A Sick or Injured Sheep |
| --- | --- |
| Is alert and curious | May be dull and uninterested in surroundings |
| When approached, will stand in a normal pose with her head up | May isolate herself from the rest of the flock |
| | May stand hunched over with head hanging down |
| | May grind teeth, which indicates pain |
| Moist, dark pink mucous membranes | Pale pink or white mucous membranes, associated with anemia and a heavy internal parasite load; blue or yellow membranes indicating disease |
| Bright, clear eyes | Dull, depressed-looking eyes |
| | A cloudy film over one or both eyes, indicating blindness or pinkeye |
| | Fresh or crusty opaque discharge in the corners of eyes |
| Dry, cool nose<br>Regular, unlabored breathing | Thick, opaque, creamy white, yellow, or greenish nasal discharge (a trace of clear nasal discharge isn't cause for concern) |
| | Wheezing, coughing, sneezing, or breathing heavily and/or erratically |
| Teeth normal for age | Broken or missing teeth usually in older sheep only |
| Mouth and tongue that are injury- and abscess-free | Sores or abscesses in mouth or on lips |
| Normal breath (remember: healthy sheep burp gas) | Putrid breath (possibly caused by abscessed teeth) |
| Blemish-free lips | Fluid-filled pimples or blisters generally indicating soremouth (see page 104) |
| Normal ears | Headshaking indicating an irritant in one or both ears, possibly parasites or weed seeds |
| Hair sheep: clean, glossy hair coats and pliable, vermin- and eruption-free skin | Dull, dry hair coat; skin showing evidence of external parasites or skin disease |
| Wool sheep: healthy-looking wool free of rubbed spots | Rubbed spots in fleece probably means keds or lice; however, some sheep and goats eat other sheep's wool, so a ragged-looking sheep might not be sick |
| Average weight for breed and age | Thin or emaciated appearance |

| A Healthy Sheep | A Sick or Injured Sheep |
| --- | --- |
| No unusual swellings on the body, legs, neck, jaw, or between the digits of their hooves | Swellings anywhere; be particularly aware of swellings over lymph nodes that could indicate caseous lymphadenitis (see page 99) |
| Free and easy movement | Slow, uneven, or limping movement; refusal to put weight on a leg |
| Healthy appetite<br>Cud chewing | Poor appetite; no cudding |
| Firm, berrylike droppings; clean tail and surrounding areas | Tail, tail area, and fleece or hair on hind legs matted with fresh or dried diarrhea |
| Clear urine | Cloudy or bloody urine |
| Soft, symmetrical, blemish-free udder with normal teats on a ewe<br><br>Normal-looking milk that tests negative using the home CMT test (see Mastitis and the CMT, page 174) | Hard, lumpy, lopsided, or discolored udder with injured or abnormally swollen teats<br>Rashes or pustules on the udder possibily indicating sore-mouth<br>Blood-streaked milk or milk containing clots of blood; watery or clumped milk |
| No vaginal discharge | Cloudy or bloody discharge from the vagina |
| Wethers and rams have a normal sheath enclosing a normal penis | Blood or crystals on the hair around the opening of the sheath; swollen sheath; dribbling urine |
| Normal temperature (see below) | A high or low temperature |

## Checking Vital Signs

. . . . . . . . . . . . . . . . . . . . . . . . . . . . . . . . . . . . . . . . . . . . . . . . .

These are normal values for adult sheep; lambs' values usually run slightly higher.

Temperature: 101 to 103°F (38.5 to 39.4°C)

Heart rate: 70 to 90 beats per minute

Respiration rate: 12 to 20 breaths per minute

Ruminal movements: 1 to 2 per minute

## Being Your Own Vet (Sometimes)

When your sheep is sick or injured, call a veterinarian. Many diseases resemble one another, such as pregnancy toxemia and milk fever, listeriosis and goat polio (sheep get it, too), tetanus and rabies, and enterotoxemia and bloat. Treating for one when your sheep has another can make the situation worse.

Many pharmaceuticals used to treat sheep are prescription drugs, some of which are off-label for sheep. Most veterinarians won't hand them out without seeing your animals. Some veterinarians will dispense prescription drugs such as epinephrine and Bo-Se to established clients, but one way or another, you'll have to get them through a licensed veterinarian.

Most veterinarians do, however, appreciate clients who can perform basic care, such as vaccinating and deworming their own sheep, treating wounds, and administering antibiotics on their veterinarian's orders.

### Using Antibiotics
· · · · · · · · · · · · · · · · ·

Antibiotic overuse is a real and rapidly growing problem. In most cases, however, strictly avoiding antibiotics isn't feasible. If your vet says to use them for one reason or another, follow directions to the letter. Use precisely the recommended dosages and complete the series as directed.

Because antibiotics destroy good bacteria as well as bad, follow antibiotic treatment with oral probiotics to restore the patient's digestive system to good health.

## Taking a Sheep's Temperature

A slightly elevated reading may be normal. Body temperatures rise slightly as the day progresses and may be up to a full degree higher on hot, sultry days. But elevated temperatures often indicate infection or dehydration; a subnormal temperature can mean hypothermia, hypocalcemia (milk fever), or possibly indicate a sheep is dying.

You'll need a rectal thermometer to take a sheep's temperature. Veterinary models are best, but a digital thermometer designed for humans works, too.

### To check the temperature:

1. Restrain the sheep. Find someone to hold him still, tie him using a halter and lead, push him against a wall and hold him there, or secure his head in the headpiece of a head gate or a milking or fitting stand.

2. Shake down the mercury if using a mercury-style veterinary thermometer.

3. Apply some lubricant (KY Jelly, Vaseline, birthing lubricants, and mineral or vegetable oils are all good lubes; your own saliva will work in a pinch) to the bulb end.

4. Hold the sheep's tail to one side. Insert the bulb end of the lubricated thermometer about 2 inches (5 cm) into a standard-size adult sheep's rectum.

5. Hold the thermometer in place, touching the inner wall of the rectum, for two minutes or until it beeps.

6. Note the temperature reading, jotting it down to report to the veterinarian. Shake down the thermometer if necessary, clean it with an alcohol wipe, and return it to its case. Store it at room temperature.

### Checking Heart Rate

The easiest way to check a sheep's heart rate is with a stethoscope, placed behind the left front elbow (in the "armpit"). If you don't have one, press two fingers between the animal's ribs behind the elbow, or against the large artery on the inside of either rear leg up near the groin. Count the number of pulses in 15 seconds and multiply by 4. Extreme heat elevates pulse and respiration, as does fear or anger.

### Assessing Respiration

Watch the sheep's rib cage, counting the number of breaths he takes in 15 seconds and multiply by 4. Be aware that heat or distress elevate pulse and respiration.

### Checking Ruminal Movements

If a sheep's rumen isn't moving normally, he is, or soon will be, sick. To check for rumen activity, place your fist in the hollow on the sheep's left side, just behind the rib cage. You should feel movement beneath your fist at least once or twice per minute.

*The color of the mucous membranes can indicate illness.*

### Checking Mucous Membranes

The thin skin that lines the inner surface of the body is called the *mucous membrane*. These membranes are good indicators of what's happening inside the body because they're so thin and transparent that you can see blood vessels through them. The easiest place to examine mucous membranes is inside the eyelid.

Check inner eyelid color in natural light. To open the eye, push the upper eyelid up with one thumb while the other thumb pulls the lower lid down. Look at the color inside the lower eyelid, keeping the eye open for a very short time.

Healthy sheep have dark pink to reddish mucous membranes. Pale pink or white membranes indicate anemia, usually caused by a heavy internal parasite load. Yellow membranes indicate jaundice, possibly due to liver disease caused by liver flukes. Dirty blue membranes are a symptom of an uncommon disease called, appropriately, *bluetongue*.

*A healthy rumen is always active and you should be able to feel the movement.*

# A Well-Equipped First-Aid Kit

Every shepherd needs a first-aid kit. Pack it in an easily carried container. We use 5-gallon food-service buckets with lids, but plastic storage or fishing tackle boxes, coolers, and backpacks work well, too. Stow it where you can find it in a hurry. A basic kit should contain:

**DISPOSABLE SYRINGES AND NEEDLES.** Syringes in 3-cc, 10-cc, and 60-cc sizes; 16-, 18-, and 20-gauge needles

**BANDAGING SUPPLIES.** Several rolls of self-stick disposable bandage material such as VetWrap, Telfa pads, gauze sponges, a roll or two of 2½-inch (6.4 cm) sterile gauze bandage, 1-inch and 2-inch (2.5 and 5 cm) rolls of paper adhesive tape, a partial roll of duct tape, and several high-absorbency sanitary napkins, which are great for applying pressure bandages to staunch bleeding

**A FLASHLIGHT.** Or better, two

**SPARE HALTERS AND A LEAD.**

**BLOOD STOP POWDER.** Flour works just as well.

**WOUND TREATMENTS.** Good ones include Schreiner's Herbal Solution, emu or tea tree oil, Betadine, Neosporin, and Blue-Kote; include topical eye ointment for injured eyes.

**CLEANING SOLUTIONS.** Bottled saline solution and alcohol, packaged alcohol wipes, Betadine Scrub

**HARDWARE.** Digital thermometer, blunt-tipped bandage scissors, regular scissors, hemostat, stethoscope, hoof pick

In addition to the basic kit you'll probably want to keep some or all of these items in your refrigerator or in a medicine cabinet in the barn:

**PROBIOTIC GEL OR PASTE.** Choose a ruminant-specific product such as Probios or FastTrack.

**BANAMINE.** A prescription pain and inflammation reliever prescribed by a vet; barring that, baby aspirin is a distant second-best for pain

**PEPTO-BISMOL.** Can be used for sheep with diarrhea

**SEVERAL TYPES OF INJECTABLE ANTIBIOTICS.** Ask your veterinarian which ones should be kept on hand.

**INJECTABLE THIAMINE.** Thiamine, also known as vitamin B1, is an important nutrition and appetite booster when sheep are sick, especially with goat polio. It's a prescription item, so get it from your vet.

**EPINEPHRINE.** A prescription injectable used to counter anaphylactic shock; can be a lifesaver

## Giving Shots

Learn to vaccinate your sheep and to give them shots. It sounds scary but is easier than you probably think. Ask your veterinarian or an experienced shepherd to show you which pharmaceuticals to use and how and where to give injections. You can buy needles, syringes, and vaccines from your veterinarian, at feed stores and farm stores, or by mail order from farm supply companies (see Resources).

Choose the correct pharmaceutical and read the label. If a product was accidentally stored incorrectly, throw it away. Also check the expiration date; don't use it if it's outdated.

Select the right disposable needle for the job. Nearly all sheep injections are given subcutaneously (SQ; under the skin) and should be given using 18- or 20-gauge needles that are ½ inch or ¾ inch (1.3 or 1.9 cm) long. A few antibiotics are given intramuscularly (IM; into a major muscle mass); in that case use a 1-inch to 1½-inch (2.5 to 3.8 cm), 18- or 20-gauge needle. Some antibiotics, such as LA-200, are very thick and the carriers used in their manufacture make these injections sting. For these, choose a 16-gauge needle so you can inject the fluid quickly before the sheep objects.

You'll need a new needle for each sheep. It's less painful if you use a new needle each time, and it eliminates the possibility of transmitting disease by way of contaminated needles.

### PREPARING THE INJECTION

**(A)** Jab a new, sterile transfer needle through the rubber cap of each bottle. *Never* poke a used needle through the cap to draw vaccine or drugs. Use the smallest disposable syringe that will do the job. They're easier to handle than big, bulky syringes, especially for women with small hands. Never try to sterilize disposable syringes; boiling compromises their integrity.

**(B)** Remove the needle from the new syringe and attach the syringe to the transfer needle sticking through the cap of the pharmaceutical bottle. Withdraw the appropriate amount of solution; detach the syringe from the transfer needle.

**(C)** Attach the needle you'll use to inject the sheep to the filled syringe. Point the needle end of the syringe upward so that any air bubbles that were created as you drew out the fluid will rise to the top. You can then press those bubbles out of the syringe.

### PREPARING AN INJECTION

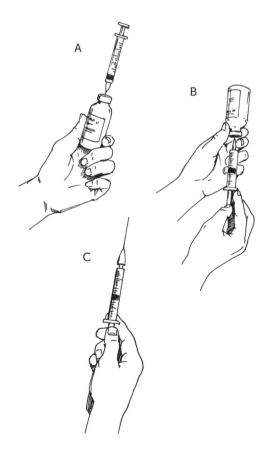

## GIVING THE INJECTION

When giving an injection, it's often easier to set a sheep up on her butt or to bend over her than to squat at the side where you're giving the injection. If you're right-handed, stand on the left side and bend over her back to the right-side injection site. When giving two injections, change sides.

Preferred sites for subcutaneous injections are the neck, over the ribs, and especially into the loose skin in the "armpit" area. Intramuscular injections are usually given into the thick muscles of the neck. When injecting a large volume of fluid (5 cc or greater for an adult sheep), break the dose into smaller increments and inject it into more than one site.

Use your fingers to separate wool down to the skin, then swab the injection site with alcohol. Never inject into damp, mud- or manure-encrusted skin.

To give a subcutaneous injection, pinch up a tent, or small fold, of skin and poke the needle

*Subcutaneous injection*

into it, parallel to the sheep's body. Be careful not to shove the needle through the tented skin and out the opposite side. Also be very careful not to prick the muscle mass below it. Slowly depress the plunger, withdraw the needle, and rub the injection site to help distribute the drug or vaccine.

Give an intramuscular injection by quickly but smoothly jabbing the needle deep into muscle mass, then aspirate (pull back on) the plunger ½ inch (1.3 cm) to be certain blood does not flow back into the syringe. If blood rushes into the syringe, it means you inadvertently hit a vein; pull the needle out, taking care not to inject any drug or vaccine as you do, and try another injection site.

### EPINEPHRINE — KEEP IT HANDY

Epinephrine, also called *adrenaline*, is a naturally occurring hormone and neurotransmitter manufactured by the adrenal glands. It's used to counteract the effects of anaphylactic shock, a serious and rapid allergic reaction that can kill.

Any time you give an injection, no matter what product or amount injected, be prepared to administer epinephrine to counteract an anaphylactic reaction. If a sheep goes into

---

### How Much Is . . . ?
· · · · · · · · · · · · · · · · ·

1 milliliter (1 ml) = 15 drops = 1 cubic centimeter (1 cc)

1 teaspoon (1 tsp) = 5 cubic centimeters (5 cc)

1 tablespoon (1 tbsp) = 15 cubic centimeters (15 cc)

2 tablespoons (2 tbsp) = 30 cubic centimeters (30 cc)

1 pint (1 pt) = 480 cubic centimeters (480 cc)

---

# Weighing Sheep

To properly administer some pharmaceuticals and nearly all dewormers, you need to know how much an individual sheep weighs.

The most accurate way to find out is to weigh him on a scale. If he's a lamb, pick him up and weigh yourself on a bathroom scale, then put him down; weigh yourself again and subtract the difference. Many veterinarians, even small-animal veterinarians, have walk-on scales at their practices; call and see if you can weigh a bigger sheep on their scale.

Another method is to tape-weigh your sheep using a cloth measuring tape. Here's how to do it:

**1.** Restrain the sheep so he's standing on the level with all four feet placed squarely under him.

**2.** Place a dressmaker's tape around his body just behind his front legs at a point slightly behind the shoulder blade, making sure it's flat and not twisted, and draw it up all the way to the skin until it's snug but not tight; this is his heart girth measurement.

**3.** Measure the length of his body from the point of his shoulder to his pin bone.

**4.** Use this formula to calculate weight:

heart girth x body length
÷ 300 = weight in pounds.

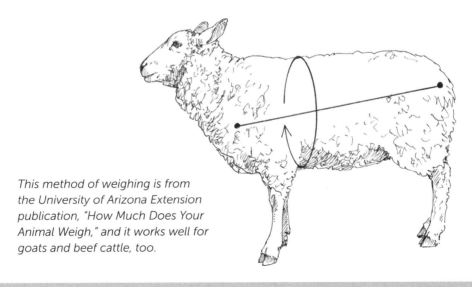

*This method of weighing is from the University of Arizona Extension publication, "How Much Does Your Animal Weigh," and it works well for goats and beef cattle, too.*

anaphylaxis (symptoms include glassy eyes, increased salivation, sudden-onset labored breathing, disorientation, trembling, staggering, or collapse), you won't have time to run to the house to grab the epinephrine. You might not even have time to fill a syringe. You have to be ready to inject it *immediately.*

If you give your own injections, buy a bottle of epinephrine from your vet and keep a dose drawn up in a syringe in the refrigerator. Kept in an airtight container such as a clean glass

Giving a syringe drench

jar with a tight-fitting lid, it keeps as long as the expiration date on the epinephrine bottle.

Take the loaded syringe with you every time you give a shot. Standard dosage is 1 cc per 100 pounds; don't overdose as it causes the heart to race.

### Drenching a Sheep

Learn to dose your sheep with liquid medicines and dewormers; this is called *drenching.* Drenches can be administered using catheter-tip syringes (not the kind you use to give shots) and turkey basters but the most efficient way is by using a dose syringe.

To drench your sheep, back him into a corner so he can't escape, restrain him (straddle his back, facing forward, if you're tall enough not to be taken for a ride), and elevate his head using one hand under his chin — not too high but let gravity help you a little bit. Insert the nozzle of the syringe between the sheep's back teeth and his cheek (this way he's less likely to aspirate fluid into his lungs) and slowly depress the plunger, giving him ample time to swallow.

When giving semisolid substances such as paste-type dewormers or gelled medications, deposit the substance as far back on his tongue as you can reach.

In either case, keep the sheep's nose slightly elevated until he visibly swallows. And watch your fingers. Sheep's back teeth are razor sharp!

### Pilling a Sheep

Ask your veterinarian if the pills he prescribes have to be given whole. If they don't, powder the pill using a mortar and pestle. Or if it's small, crush it between two spoons. If it's big, pop it in a paper bag and smash it with a hammer. Don't handle fragments or powdered pills with bare hands as some drugs can be absorbed through human skin. Next, dissolve the powder in liquid to give as a liquid drench. Or stir it into thick yogurt, or cold molasses or honey and give it like paste.

If the pill must remain whole, give it with a balling gun. This is a sheep-size plastic device designed to propel the pill down your sheep's throat. Coat the pill with cold honey, molasses, or yogurt; place it in the balling gun; and lay the assembly aside. Then secure the sheep as though you were giving him a drench. Pry his mouth open, place the loaded balling gun far enough into his mouth to deposit the pill on the back of his tongue, and raise his head to a 45-degree angle. Depress the plunger, pull out the balling gun, stroke his throat, and hope. If you're lucky, he'll swallow the pill. If he doesn't, pick it up and try again.

## Examining a Sheep's Mouth

It's very difficult to hold a sheep's mouth open to examine her teeth or to remove a piece of foreign matter from her gums or oral cavity. Sheep molars are very sharp and easily slice fingers. So, if you find you need to examine your sheep's mouth, try this method from *Sheep Management; a Handbook for the Shepherd and Student* by Frank Kleinheinz. I've tried it and it works!

"When sheep show irregularity of eating or chewing their cud, an examination of their teeth becomes necessary. The mouth of a sheep can be opened by means of two pieces of cloth, each about two feet long and two inches wide. One of these should be tied on the upper jaw, the other on the lower jaw between the front and back teeth. By pulling on these two strips one man will be able to hold the mouth open while another examines it, as shown in the accompanying illustration."

## Sheep Illnesses and Afflictions

The following are brief descriptions of fairly common health problems that you should know about. To explore them in depth, visit your favorite Internet search engine (I like Google) and type the name in the search box. Keep in mind that the accuracy of the sites that emerge from such an Internet search may vary; look for the most credible sources, such as links to veterinary schools, the USDA, and so forth. Additionally, or for those of you who do not have Internet access, invest in a good book about sheep veterinary medicine. Two especially good ones that are comprehensible to laypersons are D. G. Pugh's *Sheep and Goat Medicine* and Philip R. Scott's *Sheep Medicine* (see Resources).

### Acidosis

Acidosis, also called *lactic acidosis* or *grain overload*, occurs when rumen pH falls below pH 5.5. As pH continues to decline, rumen microbes die and ruminal action decreases or stops altogether. Symptoms include depression, dehydration, bloat, racing pulse and respiration, staggering, coma, and death. Because permanent damage to the linings of their rumens and intestines may occur, survivors generally fail to thrive.

### Dolly, a Famous Sheep

Dolly, a Finn-Dorset ewe born on July 5, 1996, at the Roslin Institute near Edinburgh, Scotland, was the first mammal to be cloned from an adult somatic cell using the process of nuclear transfer.

To create Dolly, Doctors Ian Wilmut, Keith Campbell, and their colleagues transferred an adult mammary cell from a Finn-Dorset ewe into the unfertilized egg cell of a Scottish Blackface ewe that had its nucleus removed. They caused the cells to fuse using electrical pulses. When these cells developed into an early embryo, it was implanted into her surrogate mother, another Scottish Blackface ewe. From 277 such cell fusions, 29 early embryos developed and were implanted into 13 surrogate mothers, but only one pregnancy went to full term. The 14½-pound (6.6 kg) lamb, 6LLS (alias Dolly), was born after a 148-day gestation.

Dolly lived her entire life at the Roslin Institute. When bred to a Welsh Mountain ram, she produced six lambs in three years: a single in 1998 (Bonnie), twins in 1999 (Sally and Rosie), and triplets in 2000 (Darcy, Lucy, and Cotton).

In the autumn of 2001, at the age of five, Dolly developed arthritis in a hind leg. On February 14, 2003, she was euthanized when she became ill from sheep pulmonary adenomatosis, a virus-induced lung cancer to which sheep raised indoors are prone.

When BBC News inquired about her name, Dr. Wilmut stated, "Dolly is derived from a mammary gland cell and we couldn't think of a more impressive set of glands than Dolly Parton's."

If you think your sheep has acidosis, call your veterinarian immediately. To prevent acidosis, sheep should eat mainly forage, not grain. Also be sure to make all feed changes gradually, thus allowing rumen microbes time to adapt to a different diet. Sheep such as club lambs and lactating ewes with multiple lambs who consume relatively high levels of concentrates should have access to a container of sodium bicarbonate (baking soda; buy it in bulk at the feed store) that they can nibble as the need arises.

## Bloat

Bloat occurs when a sheep overeats on grain, legume hay, or rich, high-moisture spring grass. Gas becomes trapped in the sheep's rumen and expands until it presses so hard against his diaphragm that he suffocates. A bloated sheep's sides bulge to an alarming degree; he'll breathe very rapidly and might also kick at his abdomen, grunt, cry out in pain, or grind his teeth.

If you suspect your sheep is bloated, call your veterinarian without delay. To prevent this life-threatening situation, store grain and legume hay where your sheep can't overindulge. It's also wise to feed grass hay in the morning before turning sheep onto lush, spring pasture.

If your veterinarian isn't immediately available, a homemade bloat treatment that sometimes works is made of ½ cup of vegetable oil, ½ cup of water, and 2 tablespoons of baking soda, mixed well and given with a dose syringe. Treatment with commercially available anti-bloating agents can also be tried. If you have a bloat-prone sheep (some bottle lambs fed on milk replacers, for instance, are especially prone to bloat), ask your vet to show you how to pass a stomach tube to let off gas while help is on the way; this trick may save your sheep's life.

## Caseous Lymphadenitis

Caseous lymphadenitis (commonly referred to as CL; also known as cheesy gland) is caused by the bacterium *Corynebacterium pseudotuberculosis*. CL manifests as thick-walled, cool-to-the-touch lumps containing odorless, greenish-white, cheesy-textured pus. CL abscesses form on lymph nodes and lymphoid

A CL lump; common location of CL abscesses

tissue, particularly on the neck, chest, and flanks, but also internally on the spinal cord and in the lungs, liver, abdominal cavity, kidneys, spleen, and brain. Transmission is via pus from ruptured abscesses.

Any sheep with a ripening abscess should be quarantined. The abscess should be drained into a disposable container or a latex glove and treated according to your veterinarian's instructions; for example, she will tell you where to send the sample for diagnostic testing. Because CL is transmissible to humans, wear protective clothing when draining suspicious lumps. When the procedure is completed, burn everything contaminated by pus. The sheep should remain quarantined until test results come back negative or the abscess has fully healed.

## Enterotoxemia

Enterotoxemia strain D, also known as overeating or pulpy kidney disease, is a fairly common, potentially fatal disease caused by bacteria found in manure, soil, and even the rumens of perfectly healthy sheep. Overeating on grain or milk replacer, or abrupt changes in quantity or type of feed, can cause bacteria to quickly proliferate; these bacteria produce toxins that can kill a sheep within hours.

Symptoms include standing in a rocking-horse stance and teeth-grinding (both indicate severe abdominal pain), seizures, foaming at the mouth, and coma leading to death.

Treatment is usually ineffective because death occurs so quickly.

The most common enterotoxemia vaccine is CD/T toxoid (the "CD" is for two types of Clostridium perfringens; the "T" is for teta-

nus); it's also available in an eight-way vaccine that prevents additional clostridial diseases such as blackleg. CD antitoxin provides short-term immunity to previously unvaccinated sheep.

## Goat Polio

Goat polio, also called *polioencephalomalacia*, isn't related to polio (poliomyelitis), the viral disease in humans. Goat polio is a neurological disease caused by a thiamine (B1) deficiency that culminates in brain swelling and the death of brain tissue.

Symptoms include disorientation, depression, stargazing, staggering, weaving, circling, tremors, diarrhea, apparent blindness, convulsions, and death.

If treatment begins early enough, affected sheep given thiamine injections begin improving in as little as a few hours.

Thiamine deficiencies can be caused by eating moldy hay or grain, overdosing with amprollium (Corid) when treating for coccidiosis, ingesting certain toxic plants, reactions to dewormers, and sudden changes in diet, including weaning. Overuse of antibiotics contributes to thiamine deficiencies as well.

## Hoof Rot and Hoof Scald

Hoof rot (also called *foot rot*) is caused by an interaction of two bacteria, *Bacteroides nodosus* and *Fusobacterium necrophorum*. The latter is commonly present in manure and soil wherever goats, sheep, or cattle are kept; it's only when *F. necrophorum* forms a synergistic partnership with *B. nodosus* that hoof rot occurs. *Fusobacterium necrophorum* can live in soil for only two to three weeks; it can, however, live in an infected hoof for many months. If you don't introduce it to your flock via infected sheep, cattle, or goats, you'll never have hoof rot, so take every precaution you can to avoid buying infected animals.

Sheep with hoof rot are very lame. They may hold up an infected foot and hop on three legs. If one or both forefeet are infected, they kneel to feed. Trimming infected hooves exposes putrid-smelling, pasty matter lodged between the horny outer surface of the hoof and its softer inner tissues. You'll recognize hoof rot by its stench.

Foot scald, also referred to as benign foot rot or inter-digital dermatitis, is a milder inflammation between the toes caused by *F. necrophorum* acting alone. Persistent moisture on the skin between the toes increases susceptibility to foot scald. Foot scald, however, isn't contagious, whereas hoof rot is. Sheep often get foot scald in rainy years when there's lots of squishy mud underfoot. Foot scald often precedes hoof rot when *B. nodosus* is also present.

Hoof rot is spread from infected hooves to soil to healthy animals, so isolate all infected sheep. To expose disease-causing bacteria to oxygen, trim hooves back to infected areas and remove as much rot as you can. Treat hoof rot and foot scald according to your veterinarian's instructions.

Don't buy sheep, goats, or cattle from sales barns or from infected herds. Make certain that any commercial transporters who carry your sheep thoroughly disinfect their trailers after every trip. Disinfect your shoes after visiting infected facilities. Keep pens, barnyards, and holding areas as dry as you can.

## Hypocalcemia (Milk Fever)

Hypocalcemia is caused by a drop in blood calcium. It can occur a few weeks prior to, and immediately after, lambing and is easily confused with pregnancy toxemia.

Affected ewes are depressed. They lose interest in eating and become progressively weaker until they lie down and can't get up again. Some bloat. Subnormal temperatures are common as the condition progresses.

Call your vet if you suspect hypocalcemia. The usual treatments include orally dosing with energy boosters such as glucose and the use of oral or injectable calcium substances such as calcium gluconate or CMPK (a fluid calcium, magnesium, phosphorus, and potassium product).

To prevent hypocalcemia, provide all pregnant ewes with a 2:1 calcium-to-phosphorous mineral mix at all times. It's also best to feed calcium-rich, premium quality alfalfa hay or pellets to late-gestation and lactating ewes.

## Listeriosis

Listeriosis is a type of encephalitis caused by a bacterium called *Listeria monocytogenes*. The bacteria are found in soil, plant litter, water, and even in healthy sheep's guts. Bacteria in the gut can multiply dramatically with abrupt changes in feed or weather conditions. Parasitism can trigger bacteria proliferation, too.

Symptoms include depression, stargazing, staggering, weaving, circling, one-sided facial paralysis, drooling, and a rigid neck with the head pulled back toward one flank. Symptoms resemble goat polio, rabies, and tetanus, so call your veterinarian if you suspect any of these serious diseases.

## Mastitis

Mastitis is inflammation of the udder and can be caused by a number of bacteria, including staph (Staphylococcus). Read more about it in chapter 12.

To avoid mastitis, cut back on lactating ewes' grain rations a week or so before weaning their lambs and gradually switch them from legume to grass hay. After weaning, keep them on a plain grass or grass hay diet for a week or more, until post-weaning milk buildup subsides and their udders begin decreasing in size.

## Metritis

An assisted lambing, a retained placenta, or the birth of a dead lamb should be followed by a course of antibiotics to prevent potentially fatal uterine infections, also known as *metritis*. Some veterinarians recommend inserting an antibiotic bolus directly into the uterus at lambing, some prefer antibiotic injections; ask your veterinarian in advance. Suspect metritis when a ewe becomes depressed, seems disinterested in her lambs, loses her appetite, or spends most of her time lying down a day or two to a week or so after lambing.

## Ovine Progressive Pneumonia (OPP)

Ovine progressive pneumonia is a progressively fatal viral disease. According to the online *Merck Veterinary Manual* (see Resources), the disease affects as many as 49 percent of the sheep in the western United States, to a low of 9 percent of sheep in the North Atlantic region. It's closely related to Caprine Arthritis Encephalitis (CAE) in goats.

Transmission usually occurs when lambs ingest colostrum or milk that contains the virus, or by inhalation of infected droplets in the air.

OPP is a slow-acting virus with an incubation period of one to two years. Infected sheep may eventually show symptoms of chronic pneumonia and progressive weight loss. The vast majority of OPP-infected sheep show no symptoms whatsoever; they are, however, carriers and shedders of the disease.

No vaccine and no effective treatment is available for OPP.

## Pinkeye

Pinkeye is a catchall term used to describe a number of diseases affecting the eyes; a better name is *infectious keratoconjunctivitis*. It's a highly contagious bacterial disease caused by a number of different microorganisms. It most commonly occurs during the summer months and in lambs and kids, but it can occur any time of the year and in animals of any age. Spread occurs via direct contact. The microorganisms that cause pinkeye in cattle don't infect sheep, but sheep can catch pinkeye from goats and vice versa.

The condition is painful. Infected animals try to avoid bright sunlight. Their faces below their eyes may be wet due to tearing. The eyes become cloudy or opaque and an ulcer may develop over one or both eyes. Pinkeye causes temporary blindness; in severe cases, blindness may be permanent.

Sheep with pinkeye should be quarantined in a comfortable, shady place. The disease is usually self-limiting and clears up without treatment in a week to 10 days. Most shepherds, however, prefer to clean the affected sheep's face and apply oxytetracycline ointment (Terramycin is the trade name) to his eyes three or four times a day. Sometimes the condition persists or reoccurs, in which case, consult with your vet.

## Pneumonia

Pneumonia is a common and serious disease of sheep. It's caused by several types of viral and bacterial agents and is compounded by stressful conditions such as inadequate building ventilation, poor sanitation, difficult or too-sudden weaning, abrupt weather changes, overcrowding, and shipping.

Symptoms include fever accompanied by labored breathing and a moist, painful cough. Sheep with pneumonia are depressed and may not eat. Since there are several types of pneumonia, it's important to work with a veterinarian to identify the pathogens involved and devise the most effective treatment.

## Pregnancy Toxemia

Pregnancy toxemia, also known as *twin lambs disease* and sometimes *ketosis*, is a common metabolic condition that commonly afflicts ewes during their last few weeks of pregnancy and less commonly in the first week or two after giving birth. It's mainly a condition that affects obese or thin ewes and ewes carrying two or more lambs.

Affected ewes have the odor of acetone (like nail polish remover) on their breath. They are apathetic, stop eating, and often drift away from the flock. They're reluctant to walk and spend most of their time lying down. As the condition worsens, a ewe experiencing pregnancy toxemia develops weakness in her hindquarters as well as muscle tremors, and she may collapse.

### A Cautionary Tale

A few years ago a distant neighbor showed up at our door asking if she could borrow a sheep. She'd purchased a lamb and wasn't comfortable putting her in with her goats, so she needed a companion for her until she bought more sheep. Not wanting her lamb to suffer, we picked out a gentle wether named Dimitri and took him to her farm with the promise she'd bring him home in just a few days, which she did.

We didn't quarantine Dimitri as we should have; after all, he'd only been down the road and was gone for just three days. A week or so later our sheep began getting pinkeye. It ripped through the flock and didn't respond to the usual treatments. Our veterinarian prescribed both local (squirted in the eyes) and systemic (injected) antibiotics, but some sheep went through the painful blind phase three times. In the end, our vet bill came to over $400.

When we called the neighbor to tell her our sheep had pinkeye, she said, "Oh, my goats have it. Your wether probably caught it from them. It's not a big deal; I didn't think you'd care."

The moral of this story: Don't lend sheep. And if you do, look very, very closely at the animals at the farm they'll visit. Then quarantine them on their return, no exceptions.

You can accurately diagnose pregnancy toxemia by testing the suspect ewe's urine with a test strip designed to measure ketone levels. We use a product called Ketostix; they're available at any pharmacy. To gather samples in a stress-free manner, fasten a clean can on the end of a long stick (we used duct tape to secure a baked bean can to an old mop handle; it's lasted for years). Approach the ewe while she's lying down.

Ewes pee immediately after arising, so when she stands up, be ready and casually place the can-on-a-handle behind her to catch some urine. Then, dip a strip and compare it to the chart on the package. So easy! Do, however, keep an eye on the expiration date of the test strips as they don't have a long shelf life. And don't forget to wash the can thoroughly after each use.

Pregnancy toxemia is a rapidly progressing medical emergency, so if you think your ewe has pregnancy toxemia, call your veterinarian without delay. In the meantime, offer your ewe her favorite foods (adding feed-grade powdered molasses may tempt her to eat). Also encourage her to rise, but if she can't, keep her digestive system functioning smoothly by turning her onto her opposite side every few hours.

Prevention involves maintaining pregnant ewes in moderate flesh, neither fat nor thin. Identify ewes carrying multiple fetuses and feed accordingly.

## Scrapie

Scrapie is a progressively degenerative, always fatal disease that affects the nervous system of sheep and, to a much lesser extent, goats. Like mad cow disease in cattle and chronic wasting disease in deer, scrapie is a spongiform encephalopathy caused by a prion (an infectious agent composed of faulty protein).

The name *scrapie* (pronounced scrape-y, not scrap-y) derives from the fact that the disease causes intense itching so that infected sheep compulsively scrub themselves against rocks, trees, and other solid objects, sometimes to the point of denuding themselves of much of their wool. Other clinical signs include progressive weight loss, excessive lip smacking, gait irregularities, and convulsive collapse.

A test is available but there is no prevention or cure. Scrapie is a notifiable disease, and infected sheep and their flock mates are destroyed. The United States Department of Agriculture oversees two scrapie prevention programs, one voluntary and one mandatory; if you have sheep, you *must* comply with one or the other. Talk to your County Extension agent for particulars.

## Soremouth

Soremouth, or contagious ecthyma, is caused by a pox virus that needs a break in the skin to enter the body. The disease causes blisters and scabs to form on the nose, lips, and udder of infected sheep. It's painful, so sheep with lesions on their lips are loath to eat. Ewes with lesions on their teats and udders may refuse to allow their lambs to nurse.

Soremouth is transmitted by contact with infected animals or with soremouth scabs. It can also be passed to handlers (it's called *orf* in humans), so wear gloves when handling infected sheep. The infection resolves itself in one to four weeks. Post-infection, immunity usually lasts for several years.

Don't inoculate your sheep with soremouth vaccine unless there is, or has been, an out-

Soremouth lesions

break on your farm. Soremouth vaccine is a live virus that causes soremouth at the vaccination site, thus imparting temporary immunity to the disease. When scabs on vaccine lesions fall off, however, your property will be infested with soremouth, even if it wasn't before.

## Tetanus

Tetanus occurs when wounds become infected by the bacterium *Clostridium tetani*. These bacteria thrive in anaerobic (airless) conditions found in deep puncture wounds, fresh umbilical cords, and wounds caused by docking and castration, particularly when those procedures involve the use of an elastrator. Unless treated early and aggressively, tetanus is nearly always fatal, so if you suspect a sheep has tetanus, call your veterinarian without delay.

Early symptoms include stiff gaits, mild bloat, and anxiety, progressing to standing in a rigid rocking horse stance, drooling, inability to open the mouth (hence the common name lockjaw), ear rigidity, seizures, and then death.

All lambs should be given shots of tetanus antitoxin when they're docked or castrated, and sheep of all ages should be vaccinated against tetanus.

## Urinary Calculi

Urinary calculi (also called *uroliths* or *urinary stones*) are lumps or balls of mineral crystals that form in the urinary tract. Ewes generate urinary calculi but because of their shorter, straighter urethras (the tube that empties urine from the bladder), they can pass these stones with little or no discomfort. Stones, however, easily lodge in a ram or wether's longer, skinnier urethra, especially at the S-shaped sigmoid flexure or in his urethral process, the skinny, stringlike appendage on the tip of the penis. When this happens, he can't urinate (in partial blockages he might be able to dribble), and unless the condition is quickly corrected, uremic poisoning sets in, his bladder or urethra ruptures, and he dies.

A sheep suffering from a blockage caused by urinary calculi will show at least some of these symptoms:

- Difficult, painful, and/or dribbling urination
- Blood in the urine
- Straining, tail twitching, looking back at or kicking at the abdomen
- Crying out or groaning in pain
- Standing in a rocking horse stance with front legs perpendicular to the ground and hind legs angled out in back
- Crystals in hairs around the prepuce also called the *sheath*, this is the skin covering a male sheep's penis)
- A swollen or distended penis
- Abdominal swelling from a ruptured bladder
- Lack of appetite
- Depression

Most rams and wethers who develop urinary calculi are on high-grain, low-forage diets or

# A RAM'S REPRODUCTIVE SYSTEM

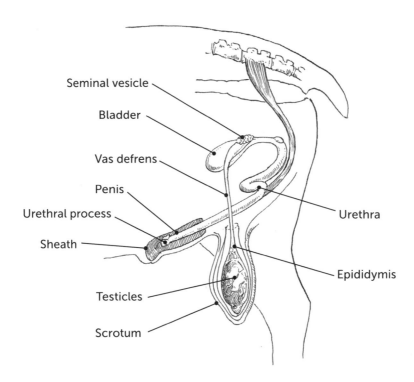

Seminal vesicle

Bladder

Vas defrens

Penis

Urethral process

Sheath

Testicles

Scrotum

Urethra

Epididymis

diets in which legume forage, especially alfalfa, predominates. Sheep rations should provide a 2:1 calcium-to-phosphorus ratio (two parts calcium to one part phosphorus). Feed rams and wethers a ration based primarily on good grass hay or pasture, along with a mineral supplement designed to balance their diets. Wethers, especially, don't need grain.

Water intake is important because it contributes to healthy urine flow. When wethers and rams don't drink enough water, their urine becomes concentrated and crystals may start to form.

Other things you can do to reduce the instance of urinary stones include:

Adding loose salt to rams' and wethers' diets at the rate of 3 to 4 percent of their total grain ration. This encourages them to drink more water.

Adding ammonium chloride to their diets to acidify their urine. Ammonium chloride makes crystal components more soluble, so they're more likely to be expelled in the urine before they form stones. Add preventive doses of ammonium chloride to grain rations at the rate of one level teaspoonful per 150 pounds of sheep, divided into two daily feedings. Ammonium chloride tastes horrible, so you may have to mix each dose with a blob of honey, molasses, jam, or flavored yogurt. Hoegger Supply

Company sells ammonium chloride by the pound.

Many veterinarians believe that early castration of lambs removes the hormonal influence needed for full development of the urinary tract. It's best to castrate future pet and fiber wethers at 4 to 6 months of age.

### White Muscle Disease

White muscle disease, also called *nutritional muscular dystrophy*, is caused by a serious deficiency of the trace mineral selenium. Most of the land east of the Mississippi and much of the Pacific Northwest is selenium deficient; these are the areas where white muscle disease is most likely to occur. Your County Extension agent is the best source of information about selenium conditions in your locale, especially in areas where selenium levels may vary widely from farm to farm.

Symptoms in lambs include weakness, inability to stand or suckle, tremors, crooked limbs, stiff joints, and neurological problems; in adult sheep, infertility, abortion, birthing problems, retained placentas, stiffness, weakness, and lethargy. Injections of Bo-Se, a prescription selenium and vitamin D supplement, sometimes dramatically reverse symptoms, especially in young lambs.

All sheep raised in selenium-deficient areas should be fed selenium-fortified feeds, have free access to selenium-added minerals, or be given Bo-Se shots under a veterinarian's direction. To prevent birthing problems and protect unborn lambs, ewes should be injected with Bo-Se four to six weeks prior to giving birth (consult your veterinarian for particulars).

**NOTE:** Make certain you are located in a selenium-deficient area before treating for white muscle disease, because too much of the mineral creates another serious condition called *selenium toxicity*.

## Coccidiosis

Coccidiosis is a potentially fatal contagious disease of sheep, particularly 3- to 6-week-old lambs; it's contracted from contaminated soil. Other livestock species are afflicted by coccidiosis, but the protozoa that cause it are species specific; your sheep can't catch dog or chicken coccidiosis.

Symptoms include watery diarrhea (scouring) that sometimes contains blood or mucous, along with listlessness, poor appetite, and abdominal pain.

Dewormers don't kill the organisms that cause coccidiosis, so sulfa drugs are the treatment of choice. Talk with your veterinarian about prevention and treatment. Sheep develop a certain amount of immunity as they age.

Two drugs, monensin (Rumensin) and decoquinate (Deccox), prevent coccidiosis, so they're sometimes added to commercial lamb feed. Rumensin is a deadly poison when ingested by horses and other equines though, so don't use that type of feed if a donkey guards your sheep.

## Internal Parasites

Sheep are troubled by four types of internal parasites: liver flukes, lungworms, roundworms, and tapeworms. All sheep have worms to some degree; it's when the worms proliferate that problems occur. The following signs may indicate an overload of parasites.

- A wormy sheep is typically unthrifty, meaning she acts as though she feels poorly and may be thin or losing weight.
- A worm problem often results in diarrhea and manure-encrusted rear ends. In severe cases, affected sheep may scour profusely.
- Young sheep are more susceptible to worms than are older sheep.
- Stressed sheep, such as late gestation and lactating ewes, newly weaned lambs, and recently transported sheep, are more susceptible to worms.
- Additional signs of worm infestation are anemia and bottle jaw (see Bottle Jaw or Milk Goiter?, page 109).

## Roundworms

Depending on their species, roundworms live in the stomach or small intestine where they feed on blood and body fluids and interfere with the digestion and absorption of feed. Roundworms are a single-host parasite; they live and reproduce in a single animal. Adult roundworms lay eggs in the stomach and intestines.

The eggs, too small to be seen with the naked eye, are passed out in manure. They hatch, progress from first- to third-stage larvae, and then crawl up blades of grass. When a sheep eats the grass, it also ingests the larvae. The larvae mature into adult worms in the intestine and stomach, and the cycle begins again.

The worst and most common roundworm, barber pole worm, is very common and potentially deadly. Badly infested sheep are thin and they usually cough. They're weak, often anemic, and most of them suffer from diarrhea. Because they're run down, they're open to infectious disease. Lambs are more susceptible to barber pole worms than are adult sheep.

The white ovaries of a female *Haemonchus contortus* twist around her red, blood-filled gut, making her resemble an old-time barber pole and earning the nematode its common name. Adult barber pole worms are big enough to be seen with the naked eye: they're ¾ inch (about 2 cm) long and about as big around as a paper clip.

A single female barber pole worm lays between five and ten thousand eggs *a day*.

One thousand barber pole worms can rob their host of up to 1½ pints of blood per day; an infestation of roughly 10,000 can kill an adult sheep.

Barber pole worms favor hot, humid conditions and they proliferate in tropical and subtropical parts of the world including the southeastern United States. Arrested larvae can overwinter in the abomasums of their hosts, making them a serious summertime threat farther north.

## Lungworms and Liver Flukes

Lungworms and liver flukes cause problems for sheep who graze on boggy pastures inhabited by snails or slugs, their intermediary hosts. Sheep infested with lungworms sometimes cough and wheeze; severe infestations cause fluid in the lungs. Low to moderate infestations of liver flukes impact growing lambs; heavy infestations can kill adult sheep.

## Tapeworms

Tapeworms, even substantial numbers of them, don't cause adult sheep much harm, but they can drastically affect the growth of lambs. Heavy tapeworm infestations, however, occasionally clog or block an adult sheep's intestinal tract. You'll know if your sheep have tapeworms if you find flat, white tapeworms or segments of tapeworms around their anuses or in their droppings. Some standard dewormers kill tapeworms, others don't. Read labels carefully to determine whether the product you're using is effective against tapeworms.

### Bottle Jaw or Milk Goiter?

Milk goiter

Bottle jaw

Bottle jaw, also called *submandibular edema*, is a puffy accumulation of fluid under the lower jaw. It usually signifies severe anemia caused by a heavy infestation of barber pole worms or liver flukes. If you see it in your flock, act quickly. A sheep with bottle jaw is at death's door.

Shepherds who have never seen milk goiter sometimes think their lambs have bottle jaw. Not so. Milk goiter is a benign, soft swelling of the thymus gland seen in healthy goat kids and hair-sheep lambs and very occasionally in wool-breed lambs. Milk goiter feels like a soft ball under the skin and occurs exactly where the throat meets the jaw. Milk goiters form as early as 1 week of age and spontaneously disappear by the time the lamb is 7 months old.

## Meningeal Worms

The meningeal worm (*Paralaphostrongylus tenius*), sometimes called the *brain worm*, affects most types of ruminants, wild and domestic, though cattle aren't known to be affected. Its natural host is the white-tailed deer, but meningeal worms' larval intermediaries are snails and slugs. Meningeal worm larvae travel up an infected sheep's spinal nerves to its spinal cord and brain. This damages the central nervous system, eventually causing death.

Suspect meningeal worm infestation if a sheep becomes fully or partially paralyzed, especially if he can't control his hind legs. Other indicators include blindness, tilting the head, circling, disinterest in food, and other signs that mimic brain diseases.

**NOTE:** Camelids such as llamas and alpacas are particularly prone to meningeal worm infestation, so if a llama guards your flock, be aware.

## External Parasites

Biting flies, gnats, and occasionally ticks tend to torment sheep, but two external parasites, lice and keds, also destroy valuable fleece. Treatment with external pesticides or dosing with ivermectin dewormers is effective; contact your veterinarian for specifics.

### Lice

Lice spend their entire lives on their hosts. Both immature and adult stages suck blood and feed on skin. They are wingless; adults are usually ¹⁄₁₆ to ⅛ inch (1.5–3 mm)long and range from pale yellowish to blue-black or brown.

Lice are generally species specific, so, for example, bird lice and cattle lice don't bother sheep; however, the same species infest sheep and goats. Louse-infested sheep itch unbearably. A sheep with lice digs his body against objects until he wears patches out of his fleece or hair coat. A louse-infested sheep is generally listless, and in severe cases, loss of blood to sucking lice leads to serious anemia. Infested sheep may stop eating and lose weight.

There are two types of lice: sucking lice that pierce their host's skin and suck blood and biting lice that have chewing mouthparts and feed on particles of hair and scabs. Lice are usually spread by means of direct contact, often when infested sheep join an existing flock. Lice proliferate during autumn and reach peak numbers during late winter or early spring. Summertime infestations are rare. Wintertime infestations are usually the most severe.

If one sheep requires treatment for lice, treat the whole flock, keeping in mind that louse control is difficult because pesticides kill lice but not their eggs. Since eggs of most species hatch eight to twelve days after pesticide application, re-treatment is necessary two or three weeks following the first application.

## Keds

Sheep keds (*Melophagus ovinus*) are ¼-inch (6 mm) long, flat, leathery, wingless flies that resemble ticks. Unlike true ticks, keds spend their entire life cycle on a sheep, although they can spread by crawling from one sheep to another sheep. Adult sheep keds live up to 6 months, during which time a female produces about 15 young at the rate of approximately one each week.

Keds crawl over their hosts' skin and feed by inserting their sharp mouthparts into capillaries and sucking blood, much like a mosquito. This causes considerable irritation, which causes the host sheep to rub, bite, and scratch at its wool, reducing the amount and quality of his fleece. Feeding punctures also cause a condition known as *cockle* in tanned skins. Large ked infestations cause anemia, which weakens the sheep and makes him more susceptible to other diseases.

## Fly Strike

Fly strike, also called *myiasis*, is caused by maggots devouring a living sheep's flesh. It isn't the problem in North America that it is abroad because we aren't plagued by many of the species that do the most harm. We do, however, have pockets of blowflies (bluebottle and greenbottle flies) and screwworm flies in the United States, so it pays to know what to look for.

Open wounds attract flies. Add wet or diarrhea-soaked wool to the equation and fly strike is apt to occur. When blowflies and screwworm flies lay their eggs in wounds, the eggs hatch within hours. Larvae (maggots) secrete enzymes that liquefy the skin and flesh of the animal on which they feed, so untreated wounds quickly become septic and spread. Untreated sheep can die from fly strike, so carefully examining all sheep at least twice a day in areas where blowflies or screwworm flies are present is important. If you aren't sure about what resides in your locale, ask your County Extension agent.

## Watch Where You Put That Dewormer

A few years ago we dewormed our sheep with Levasole, the kind in powder form to be mixed with water and given as a drench. When we were finished there was some left over. John took it in the house and I forgot about it.

Background: we both got hooked on mineral water when we lived in Minnesota but the only kind we can buy in the Ozarks is imported Perrier. Considering the cost, we save it for special treats. Perrier comes in sturdy, dark green, glass bottles that I refill with home-filtered water and a squeeze of fresh lemon juice for an everyday refresher.

The day after deworming the sheep, I reached into the refrigerator, grabbed a Perrier bottle, and glugged a huge slug. "Ack!" I thought, "That lemon has gone *bad*!" Then I looked closer and in tiny lettering on the label, John had written "Levasole."

I went to the computer and Googled Levasole, then hied myself to the bathroom, tossed a handful of salt down my throat (because I have an incredibly hard time making myself throw up), and hugged the porcelain throne for the next ten minutes.

Lesson: never, never, *never* store leftover dewormers or pharmaceuticals in food containers, especially containers you might dig into without looking at them too closely. It goes without saying that John will never live this down.

# Part 3
# Making More Sheep

*These two early twentieth-century children obviously didn't know better than to ride their very patient sheep.*

CHAPTER **8**

# Breeding Sheep

*Little Lamb, who made thee?*
*Dost thou know who made thee?*

— William Blake, *"The Lamb"* (1789)

Many shepherds say, "Don't own a ram unless you need one." If lambs are in your future, you do need one. Otherwise, you'll have to borrow a ram or take your ewes to a ram to be bred. In most instances it's better to keep your own ram for breeding.

## Owning Your Own Ram

If you have more than two or three ewes, and there are no compelling reasons to avoid keeping a ram (for example, lack of facilities or the presence of small children or vulnerable adults in your household who might wander into the ram's pen and get hurt), it's much less bother to own your own ram than to arrange for the use of someone else's.

If he's yours, you'll know he's healthy and that he won't bring hoof rot or other diseases to your flock like a loaner ram might do. You'll also be assured that your ewes won't pick up any diseases as they might if they were to go elsewhere to visit a ram.

Buy (or, if need be, use or rent) a quality ram who complements your ewes. Every breeding ram should exemplify the qualities you're breeding for, be it type, wool, size, meatiness, or any other characteristic. He should be healthy,

### My Take on Rams

· · · · · · · · · · · · · · · ·

I like rams. We have six of our own plus two long-term boarders. They aren't terribly destructive and none are aggressive toward humans. Still, we carry our shepherd's crook and keep an eye on the boys when we enter their paddock. Even docile rams sometimes react out of character. They're sweet, but we never take them for granted.

free of specific breed faults (such as split eyelids in Jacob sheep or horns that spiral too close to the face in heavily horned breeds such as Icelandics and Scottish Blackface), and have two large testicles. Size of the testicles equates with fertility in rams. A truly bad-tempered ram is never a bargain; it's not much fun interacting with your flock when you have to constantly watch your back.

## Buy a Proven Ram

Eight percent of domestic rams prefer rams or wethers instead of ewes as sexual partners, according to research from the Oregon Health and Science University School of Medicine. Scientists don't believe this behavior is related to dominance or flock hierarchy; instead, it seems to be associated with a region in the

---

## What Happens When

Use these guidelines to plot your breeding strategies, keeping in mind that sheep sometimes don't read the book. Precocious 3-month-old ram lambs have, for instance, impregnated their sisters and their dams.

### Ewes

**Age at puberty:** 5 to 10 months (single ewe lambs cycle younger than do lambs from multiple births)

**Earliest breeding of spring ewe lambs:** when they reach 60 to 75 percent of their expected adult weight

**Heat cycle:** every 13 to 19 days

**Heat duration:** 24 to 48 hours

**Ovulation occurs:** 20 to 30 hours after onset of estrus

**Embryonic implantation occurs:** 21 to 30 days after fertilization

**Length of gestation:** average 147 days, normal range 138 to 159 days (averages vary slightly by breed; primitive and early-maturing breeds generally have shorter gestations)

**Number of young:** 1 to 7 (singles or twins are the norm for most breeds)

**Length of breeding season for seasonal breeds:** August through February

**Length of breeding season for aseasonal breeds:** year-round but strongest August through February

### Rams

**Age at puberty:** average 5 to 7 months or 50 to 60 percent of mature weight (some breeds reach puberty much younger than this)

**Primary rut occurs:** August to January (though most rams will breed ewes year-round)

**Breeding ratio:** 1 adult ram to 35 to 50 ewes; 1 ram lamb to 15 to 30 ewes

---

rams' brains that is half the size of the corresponding region in heterosexual rams. To avoid problems, it's best to buy a proven ram. Or, if you choose a ram lamb for your future sire, ask for a written breeding guarantee.

## Using Someone Else's Ram

Be careful if you take your ewes to another person's ram. Be certain the destination farm is free of hoof rot and CL and that your ewes will be housed separately from the farm's other sheep. If you're taking a single ewe, find out where she'll be kept; she should have her own quarters but be able to see other sheep. Find out in advance what the breeding fee covers: Breeding through a single heat cycle? Return breedings until she becomes pregnant? Will you be charged extra for board? Can you bring your ewes' accustomed feed if you want to? Work out the details (and put them in writing) before you commit.

If you rent or borrow a ram, establish who is responsible for veterinary bills if he becomes injured as well as what happens if he dies in your possession. Allow enough time before breeding season begins to quarantine him for 30 days, even if you're sure he comes from a disease-free flock.

## Handling Rams

Most wool sheep are seasonal breeders, which means they're most likely to breed in the fall, when days are growing shorter and temperatures cooler. Ewes become receptive and rams go into rut. This is when you must take special care around your ram; even gentle rams get edgy this time of year.

The verb "to ram" is derived from the Old English word *ramm*, meaning an intact male sheep. And ramming is what rutting rams do. They back off a distance, lower their heads, then race forward and smack their opponent, poll first. Crash! Some rams show aggression toward humans by backing, running, then jumping and slamming to a halt just before connecting with their human target. Don't allow it!

If your ram is only occasionally feisty, be ready with a bucketful of water to dash in his face. Or, if you're quick, strong, and fearless, grab him mid-charge and roll him over, then hold him down for at least five minutes.

Most butting behavior toward humans can be avoided by never fooling with a ram's face or forehead, beginning when he's a tiny lamb. Never play pushing games with a ram lamb; it's cute, but he'll grow up thinking you're his opponent. It's not as amusing when he's grown and takes you by surprise.

If cornered by an ornery ram, look big. Stand tall and hold your arms out to the sides. Extend your outward reach using sticks, a shepherd's crook, or anything else at hand. Don't turn your back. The ram should back down first. If he charges, rap him smartly on the fleshy part of his nose with your stick. Aim well. Don't injure his eyes or smack his forehead, horns, or poll. If you do, he'll think you're extending a challenge.

To reward your ram, scratch his chin. He'll love it and it encourages him to raise his chin, which in turn defuses his inborn instinct to lower his head and charge. You can also scratch

*A ram shield prevents a rutting ram from charging and butting.*

his back or his chest, but only if he's gentle and not in rut. Leave his head alone!

And if he's a good ram most of the time, fit your seasonally testy ram with a ram shield. This is a sturdy leather mask that allows him to see to both sides but not home in on a frontal target. Sheep, however, sometimes gang up on an individual wearing a ram shield, so remove aggressive flock mates if they won't leave him alone.

## Bell the Ram

A good way to keep track of a potentially ornery ram is to hang a bell on a collar around his neck. Then you can tell where he is at all times. It's best, however, to use a bell collar that will break if he snags it on something really solid; you don't want him to hang himself or break his neck.

## Dances with Sheep

**In 2002 we moved** from our woodland acres in east central Minnesota to a ridgetop farm in the southern Ozarks. One of the first things we learned was that our neighbor across the road raised cattle, lots of cattle. In addition to his own property, he rented six hundred acres of pasture from another neighbor down the road whose acreage abuts ours in back. He and the two cowboys who worked for him frequently rode down the dirt-track township road that fronts our farm, either on horseback or driving a battered farm truck.

A few weeks after our arrival, I spied a large spider sunning itself on the paved county road. A (mostly) recovered arachnophobe, I was determined to see it close at hand, so I pulled to the roadside and sidled up to the creature. As I stood there clenching my teeth, clutching my fists to my chest, and wondering if I dared get any closer, the neighbor's battered pickup came barreling down the road. It screeched to a stop, and the cowboy on the passenger's side asked me, "You got trouble?"

"No," I told him, "but thank you. I was just wondering what this is." (Meaning, "Is this an Arkansas Chocolate Tarantula?")

The two cowboys exchanged a knowing look. ("Really ignorant northern woman," was what it said.)

"It's a spaahder [spider]," he told me.

After that I became a continuing source of amusement for them, or so it seemed to me. Whenever I did something stupid, they were there. Like the time I chased our errant, runaway beagle down the driveway and out into the road, wearing my jammies, just as the battered pickup crested the hill. Or the hot summer's day I was kicking back, eyes closed, luxuriously cooling off in the llamas' wading pool, and they came trotting up our driveway on horseback to borrow a dose of Banamine.

In December 2003 I bought my first registered Miniature Cheviot sheep, Baasha, and a ram lamb named Abram. Abram was ultra-friendly, and he quickly became a pet. A bad idea, I knew, but he was so darned cute.

(Background: I learned about woollies by helping a friend with her small flock of spotted Jacob sheep and her very ornery ram. It wasn't his fault; as a lamb his owners played with him by pushing on his forehead so that he'd push back.

When he grew up and began ramming people in earnest, the family returned him to his breeder, who sold him to my friend. After I was knocked down and thoroughly worked over by this ram, my friend showed me how to subdue him by grabbing a front leg, moving quickly to his side, and flopping him down on the ground.)

So, back to Abram: As I walked among the sheep (in the pasture bordering the township road, of course), he would flip into ram mode and begin backing up to charge. Then he'd race toward me, hopping to a stop just before making contact. I didn't like that, so I took my friend's approach and tried to flip him to the ground. Miniature Cheviots, however, have long, broad bodies set on short, wide-set legs, so I couldn't quickly roll him over as I'd learned.

Instead I grabbed both front legs; by lifting his fore end and throwing my body against his shoulder, I could lay him down. Exactly twice I could do this. Then he discovered that if he hopped on his hind legs, I couldn't get beside him to shoulder him down.

One morning after Abram discovered the hopping trick, we were waltzing across the field when a familiar battered truck crested the hill. It slammed to a halt, and its grinning occupants watched us with open delight. I later learned (through a mutual acquaintance who was there) that they were on their way to a café in Hardy, where they entertained the coffee-drinkers' table with loud guffaws and a vivid description of the crazy lady up the ridge who dances with sheep.

Not long after that our neighbor downsized his cattle operation and the cowboys stopped driving by. I still do stupid things, but in a perverse way, I sort of miss my audience.

## The Sheep-y Birds and Bees

The two basic modes of breeding are *pasture breeding* (the ram lives in the flock with the ewes) and *pen breeding* (both parties are taken to a pen where the mating occurs). Since we have multiple rams, we prefer pen breeding. Since our breeding pen is situated where we can keep an eye on what's going on, we know the exact date when each ewe is bred.

Ewes in heat seek out and stay near the ram. Some sniff, lick, or nuzzle him as well. A ram approaching a potential girlfriend sniffs her urine, nudges her, grunts, pants, and flehmens. If the ewe is in heat, she'll stand for his advances; she may also wag her tail. If disinterested, she'll simply walk away.

Mating is achieved very quickly. Unless you're observant you probably won't see it occur, especially in a ram-in-the-pasture breeding situation.

## Handling Ewes Post-Breeding

Embryos aren't implanted in a ewe's uterus until 21 to 30 days after she becomes pregnant, so during that time avoid shearing, hoof trimming, showing, shipping, or anything else that might upset recently bred ewes.

Check vaccine and dewormer labels and if they shouldn't be administered to pregnant ewes, use a wide-nib felt-tip marker to prominently note the fact on each bottle. I once absentmindedly dewormed my flock with Valbazen, even though I knew it shouldn't be given to ewes within 45 days of conception (some shepherds, myself included, recommend avoiding it throughout pregnancy), and all of my lovely ewes aborted their lambs. Learn from my mistake: marking bottles with eye-catching warnings sometimes saves a world of heartache later on.

## Feeding Pregnant Ewes

Feeding pregnant ewes is an art. Early and mid-gestation are critical periods because placental development occurs between 30 and 90 days after conception, while 70 percent of fetal development takes place during the last 4 to 6 weeks of gestation. But pouring on the feed isn't the answer; overweight ewes are prone to pregnancy toxemia (see chapter 7), vaginal prolapses, and dystocia (lambing problems). What and how much to feed depends on many factors including your ewe's age and breed, her body condition, how long she's been pregnant, how many fetuses she's carrying, when her lambs will be born, and types of available feed.

Early- and mid-gestation ewes should be fed a quality maintenance diet (see chapter 6). They shouldn't be thin, but don't let them become obese. They don't require concentrates, but they do need good pasture or hay and access to sheep-specific minerals formulated for farms in your area.

### Late-Gestation Feeding

Late-gestation ewes require additional energy, protein, calcium, selenium, and vitamin E. Energy equates with concentrates (grain). Adjusting for her pre-pregnancy size, the real or estimated number of lambs she's carrying, and the quality and quantity of pasture or hay available, begin adding a small amount of grain to your ewe's diet about six weeks prior to her due date, working up to 1 pound (450 g) per day just prior to lambing. According to the *Merck Veterinary Manual,* it also helps to include niacin in late-gestation ewes' diet at the rate of 1 gram per day.

An exception to the rule is for ewes that you know, or strongly suspect, are carrying three or more lambs. As their fetuses mature, these ewes run out of room to properly digest hay, so you'll want to limit their roughage intake. Instead, feed small amounts of high-quality hay, such as prime alfalfa, adding 1 pound (450 g) of grain per day per fetus, split into at least two feedings per day.

And because they're still growing themselves, late-gestation spring ewe lambs require more energy by way of concentrates. Work them up to 1 pound (450 g) of grain twice a day.

Ewes' requirements for calcium more than double during late gestation. Grains are notoriously low in calcium, though some high-quality forages, especially legumes such as alfalfa and clover, can supply most ewes' needs. Supplying additional calcium through commercial grain mixes formulated for pregnant ewes (these also supply additional protein) or through high-calcium mineral supplements is a better approach.

Additional selenium and vitamin E are also needed during late gestation. Low selenium levels are implicated in retained placentas in ewes and white muscle disease in lambs. It's important to know if your pastures or the land on which your feed is grown is selenium deficient. If it is, feed loose minerals that are high in selenium and vitamin E year-round or give late-gestation injections of a prescription substance called *Bo-Se.* Because we live in a very selenium-deficient zone, we give our pregnant ewes and does a Bo-Se shot about five weeks prior to lambing or kidding, at the rate of 1 cc per 40 pounds (18 kg) of body weight. However, too much selenium is toxic, so it's very important to discuss its use and dosage with your County Extension service or veterinarian.

## How Many Lambs?

The trick to feeding late-gestation ewes is knowing how many lambs they're carrying. The problem is, how do you know?

Ask if any local veterinarians offer pregnancy testing by means of ultrasound. An adept technician can ultrasound sheep between 25 and

90 days of pregnancy and tell you how many lambs each ewe conceived. Easy!

Otherwise, assume that exceptionally bulky-looking late-gestation ewes, especially of breeds known for multiples, are carrying more than one lamb; it's usually better to judiciously feed for multiples and have a single lamb than the other way around.

## Late-Gestation Tips

Your late-gestation ewe needs plenty of stress-free exercise. A good way to help her move around more is to place feeders, water containers, and mineral tubs or dispensers in separate areas of pastures and paddocks so she has to walk back and forth between them. Give her a CD/T toxoid booster between five and seven weeks prior to lambing. She will then pass immunity to her lamb through her antibody-rich colostrum (first milk). If you live in a selenium-deficient area, this is the time for her Bo-Se shot as well.

Although it's wise to forgo stressful situations as your ewe's due date approaches, shearing prior to lambing is a good idea. With her fleece gone you can easily monitor her udder development, and her lambs will have an easier time finding her teats. There are two more good reasons to shear just before lambing.

Stress, and lambing is very stressful, usually causes a wool break (a weak spot) at skin level in a ewe's fleece. If she's shorn just before the break occurs, the break won't devalue her fleece. Also, a fully fleeced ewe with new lambs is much more likely to take them out in the cold and wet than will a newly shorn ewe. Use common sense, though, and don't shear winter-lambing ewes if the temperature is likely to fall astronomically low.

If shearing is out of the question, crutch your ewe about two weeks before lambing; crutching (sometimes called *crotching*) consists of trimming wool and dags (manure and urine-soaked locks of fleece) from around her udder and hind end. We use sharp Fiskars scissors for this job to reduce the sort of stress that electric shearing by amateurs can cause. And if you set her up on her butt, work quickly; a late-gestation ewe has trouble breathing in a tipped position. This is also the time to deworm her and to trim her feet.

**NOTE:** It's especially important to monitor late-gestation ewes for pregnancy toxemia and hypocalcemia (milk fever), which are two common, potentially fatal metabolic diseases. See chapter 7 for a complete discussion.

## What to Do Before Lambing Starts

I can't imagine anything more exciting and rewarding than lambing time! Lambing isn't, however, a stroll in the park. Proper nutrition and management can mean the difference between life and death for your ewe and her lambs. And if at all possible, you *must* be there when she lambs (you might want to use a baby or barn monitor at night). She might need help.

People say, "Sheep have been giving birth to lambs unassisted since time began." That's true, but a lot of ewes and lambs have needlessly died. In most instances, lambing goes off without a hitch, but when a ewe or lamb needs help, they need it badly.

## Provide the Right Place

Ewes like to pick their own place to lamb, but it should be in a clean, dry environment. To reduce stress (sheep don't handle change well), move your ewe to the lambing area at least a week before her first due date. She needs company, so take along at least two other sheep, preferably her friends or daughters or her mom.

A grassy paddock with shelter is ideal in good weather; otherwise, a large, clean area indoors will do. It shouldn't be small. A ewe moves around during first-stage labor; lying down, getting up, and lying down again helps position her lambs for a safe delivery. Don't pen her in a stall; she needs room to maneuver.

When she picks a spot, don't move her unless there is a compelling reason to do so. Lambing in a light, warm rain is perfectly okay; outside in torrential rain or a blizzard is not.

Once she enters second-stage labor, she should stay where she is until her final lamb is born. Don't move her to a jug (a mothering pen) until she's finished. Moving her to a jug mid-delivery disorients her, interrupts her labor, and gives her less room to lamb. And, in such tight quarters, she's much more likely to lie down on a newborn lamb while giving birth to its siblings.

A note about moving her to a jug after lambing: if you pick up her lamb or lambs and use them to show the way, she'll follow. Ewes are not, however, wired to look for flying lambs. You must bend over and carry them with their feet barely skimming the ground. This is a backbreaker and unwieldy if more than two lambs need to be moved. Save your back; buy a lamb sling. Also, keep in mind that if she loses sight of the lambs for an instant, she will race back, frantically *baah*-ing, to the spot where she gave birth to look for them. There always seem to be a few false starts, so expect them.

## Set Up a Shepherd's Room

Frank Kleinheinz's advice (see page 125) is as good today as it was when he wrote it 100 years ago. It's important to be there when ewes lamb, whether you have two ewes or two hundred.

Years ago when I helped foal mares for a horse breeder friend, I perfected the art of sleeping in a barn. It's an adventure! First, find the perfect spot for your sleeping and observation area. You should be situated where you can see your ewe but not so close that your presence makes her nervous.

Choose a comfortable cot or make a comfy bed by stacking a row of hay or straw bales side by side and long enough to accommodate your frame. Don't use blankets; they pick up debris and require rearrangement every time you get up. Instead, choose a sleeping bag you can easily slip into and that is rated for the depth of cold you'll face during lambing time. Avoid bags with flannel lining; these pick up debris, too.

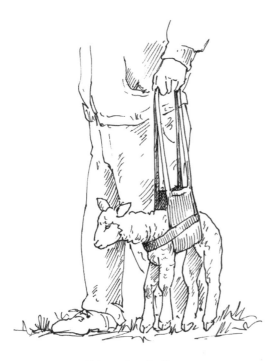

Using a lamb sling

Lighting can be a low-watt bulb left burning from dusk to dawn or (my choice) a string or two of Christmas lights; they throw enough subdued light to keep tabs on ewes and lend a festive air. If you're far from the house and have a safe place to put it, install a microwave for heating water (and snacks, should you get the munchies in the middle of the night).

Don't play music unless your ewe is accustomed to it. Reading is good, but bring an extra flashlight so you don't run down the batteries of the flashlight in your lambing kit. In dim overhead lighting, a flashlight can help you see more clearly exactly where you need extra light. Check the ewe often, being as unobtrusive as you can.

## Be Prepared

Most ewes lamb without assistance, but you should be ready to help if the need arises. As your ewe's due date approaches, clip your nails short and smooth the edges; you won't have time for a manicure if you have to assist.

Program appropriate numbers into your cell phone; be sure you can phone a vet or a shepherd friend or two at the touch of a button.

Assemble a lambing kit and the supplies you'll need if you find yourself with an orphan to feed. Collect nipples, bottles, and tube-feeding apparatus (see Lambing Kit, page 127), and if you don't have a lactating dairy ewe or milk goat to provide milk for unexpected bottle babies, buy a bag of quality milk replacer to have on hand.

Ewes and their lambs should be placed in individual jugs for two to four days after lambing, so have one or more set up in advance. Jugs are small, safe pens where a ewe can bond with her lambs and recover in peace. Older, experienced ewes with single lambs might not need jugging but don't count on it. It's best to use a jug every time.

Set up jugs where drafts won't be a problem. If that's not possible, use jugs with solid sides. Otherwise, sheep panels set with their smallest openings near the ground and cut to size (4 x 4 feet [1.2 x 1.2 m] for ewes of small breeds such as Shetlands and Soay; 4 x 6 feet [1.2 x 1.8 m] or 5 x 5 feet [1.5 x 1.5 m] for bigger breeds) and secured with snaps or wire work very well.

Or improvise. We've used our two-horse trailer divided with safely stacked bales of straw when several ewes lambed at the same time and we needed one more jug.

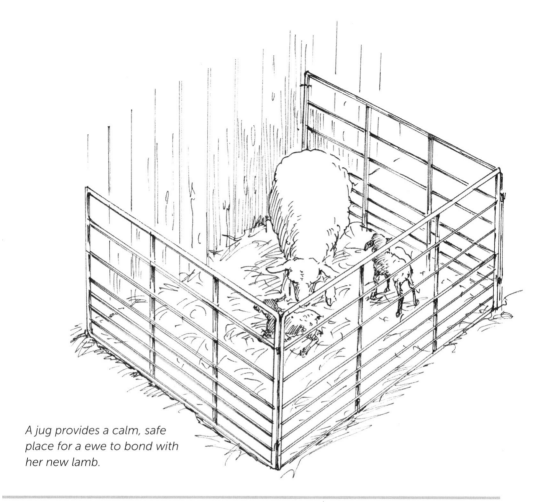

*A jug provides a calm, safe place for a ewe to bond with her new lamb.*

**LEARN FROM THE PAST**

On account of the dangers of lambing time it is most essential that the shepherd be near the flock at all times during this period. As a good shepherd must give up many hours of sleep in order to raise as large a percentage of lambs as possible, a small room should be provided for him in the sheep barn close to the lambing pens so that he may be comfortable during his weary watch. In this room should be a cot or bed upon which he can lie down when his duty does not require him to be with his flock. A stove should also be furnished so the shepherd may keep warm in cold weather. By keeping a teakettle of water on the stove he will always have warm water on hand, which is often needed. Otherwise, if he should find a chilled lamb which needs a warm bath at once to revive it, he will be compelled to run to the house, build a fire, and warm water, causing serious delay.

— Frank Kleinheinz, *Sheep Management: A Handbook for the Shepherd and Student* (1912)

Jugs shouldn't be set up over cold, damp concrete floors, which even when bedded can chill newborn lambs. Bed with clean, non-musty straw or hay, not with wood chips or sawdust that can irritate lambs' tender respiratory tracts.

Give water to jugged ewes in shallow containers. Never use deep buckets of water in a jug lest lambs fall in and drown.

## Is It Time?

A first-timer's udder may begin enlarging as much as two or three weeks before lambing; a veteran ewe's udder starts filling anywhere from ten days to a few hours before delivering her lambs. Most ewes develop strutted udders a day or so before lambing. A strutted udder is so engorged with colostrum that it's firm and shiny and the teats jut out somewhat.

A ewe's perineum (the hairless area around her vulva) usually bulges during the last few weeks of pregnancy. About 24 hours before lambing, the bulge subsides and the vulva becomes longer, puffier, and increasingly flaccid. The udders and vulvas of pink-skinned ewes flush a much deeper pink.

Release of the hormone relaxin causes structures in a ewe's pelvis to soften as lambing approaches. As this occurs her rump becomes increasingly steeper; the area along her spine sinks and her tail head rises. If she isn't shorn but she's tame, begin running your hands over the ewe's body every day starting a few weeks before she's due; then you can feel, rather than see, these changes.

# Lambing Kit

Store your lambing supplies in something easy to lug around. We used to keep ours in a Rubbermaid toolbox-stool. It was roomy, it had a lift-out tray for small items, it was easy to carry to the barn, and it was much more comfortable to sit on than an over-turned five-gallon bucket. Then the handle broke off, so we replaced it with a Coleman cooler. It's even roomier and nicer to sit on! A large fishing tackle box is another fine option, or a book bag or day pack to sling on your back. You will need:

A bottle of 7% iodine to use to dip newborns' navels. This is now a prescription item, so unless your veterinarian carries it, you might have to substitute a product like Triodine-7 (a weaker iodine of tincture) or Nolvasan (chlorhexidine); ask your vet for recommendations.

Shot glass or plastic film container to hold navel-dipping fluid

Scissors for trimming extra-long umbilical cords prior to dipping; disinfect and store them in a sturdy resealable plastic bag

Dental floss to tie off a bleeding umbilical cord (rarely needed)

Digital rectal thermometer

Bulb syringe, which are the kind used to suck mucous out of human infants' nostrils

Shoulder-length obstetrical gloves, if you like them (I assist bare-handed)

Plentiful supply of obstetrical lubricant; you can't have too much. We use SuperLube from Premier1.

Antiseptic cleanser like Betadine Scrub for cleaning ewes' vulvas and human arms prior to assisting

Sharp pocket knife, because you never know when you might need one

Hemostat (ditto)

Lamb-carrying sling

Halter or collar and lead

Reliable flashlight with extra batteries

Clean terrycloth towels or a large roll of heavy-duty paper towels

Two or three lamb coats

Lamb feeding equipment (see pages 144—149)

**A NOTE ON LAMBING SNARES:** I've heard so many stories about people tearing ewes with the cable and plastic kind that I would never recommend using one. The rubber ones are absolutely useless. I have a length of plain, clean, soft rope with slip loops tied on both ends that I can use if necessary and then discard and replace it.

# We Have Lambs!

**As a young woman** I loved James Herriot's books. Sometimes the stories almost made my hair stand on end as the good doctor labored to deliver stuck lambs and calves. The thought, "I could *never* do that!" kept me from getting sheep for a very long time.

Eventually, I helped many foals be born, from my mares and from mares belonging to a breeder friend. She also had sheep, and after watching several uncomplicated lambings, I figured I could handle that, too. Fate unknowingly smiled on me when I bought my first sheep.

My Classic/Miniature Cheviots are an old British hill breed selected for hardiness, easy lambing, and strong mothering instincts. But I do occasionally have to lend a helping hand, so as lambing approaches I make sure there's plenty of lube in the lambing kit, and I remove my rings and file my fingernails short, just in case.

Late one afternoon, Wren, one of my favorite ewes, got that faraway look in her eye, an early sign of first-stage labor. Then she took herself to the far side of the pen and began pawing up dirt to make a nest. She'd stop, listen, turn around, and look at the ground ("Where are those lambs?") and paw again. I brought my lambing kit, a book, and a can of soda out to the pen and settled in to wait.

By dusk she'd crept into the straw-bedded stall I planned to partition off for a mothering pen, but she was bothered by Nick, a wether I kept with her for company. Nick is kind of a big dumb boy, so he followed her around and kept getting in the way. I locked him out of the stall, and Wren got down to business.

I watched through the gate until the water bag appeared at her vulva. When she'd settled into her nest and stayed down and started straining, I quietly crept up behind her. Two front feet followed by a nose. Perfect! I encouraged ("Here comes lambie!"), Wren pushed, and out slid a perfect black lamb.

I squeegeed fluid from the lamb's face with my fingers, and when Wren got up and the umbilical cord ruptured, I placed the lamb by her face. But first I took a peek. A ewe lamb! Yay! All of last year's lambs were boys, so I'd put in an "order" for ewe lambs.

Wren crooned to her lamb and licked her. And licked and licked and licked. A short while later Wren's head shot up and she looked surprised. She stopped licking, slung herself down, and popped out another beautiful, black ewe lamb.

Although I've assisted à la Herriot several times, things usually go according to plan, and it isn't as scary as I thought. I love lambing time!

# CHAPTER 9

# Lambing Season

*Thy teeth are like a flock of sheep that are even shorn, which came up from the washing; whereof every one bear twins, and none is barren among them.*

— Song of Solomon (4:2; King James version)

bout 95 percent of lambs are born in the normal nose and front feet first diving position. Furthermore, a normal delivery usually takes about five hours from the start of cervical dilation to delivery of the first lamb: four hours for cervical dilation and one hour for expulsion of the lamb.

Ewe lambs giving birth for the first time, obese ewes, skinny ewes, old ewes, and flabby ewes who didn't get enough exercise in late gestation are more prone to birthing problems than are young, physically fit ewes.

In multiple births, the same sequence of rupture of the water bag and birth of a lamb is repeated for each delivery. If a second water bag appears at a ewe's vulva as she's cleaning one lamb, another is on its way. A normal interval between deliveries ranges from 10 to 60 minutes.

Lambing can be divided into three phases: first stage, when the ewe's cervix dilates; second stage, the birth of her lambs; and third stage, expulsion of the placenta.

## First-Stage Labor

During first-stage labor, mild to moderate uterine contractions may cause a ewe to stretch and raise her tail or to lie down briefly and hold her breath. The following behaviors may also indicate first-stage labor, which generally lasts 4 to 12 hours.

- She will probably drift away from the flock to seek a private birthing spot.
- She'll "nest" by digging a depression, lying down, getting up, circling, digging, and repeating the cycle over and over again.
- She'll walk more loosely on her hind legs (as hormonal changes relax her pelvis, her rear leg angulation changes, which affects her gait) and move more slowly than usual.

- During early first-stage labor she may pant, flehmen, swish her tail, *baa* loudly, or grind her teeth (all indications of pain) as abdominal contractions lasting 15 to 30 seconds occur at 15-minute intervals. Contractions occur more frequently as first-stage labor progresses, until they're 2 to 3 minutes apart.
- As first-stage labor progresses her sides will seem flatter and her belly lower as her lambs move into the birth canal.
- As her cervix dilates, a string of mucous composed of the cervical plug will likely slide from her vulva.
- She may frequently crouch to urinate or attempt to urinate.
- Some ewes search for their unborn lambs while vocalizing in soft, low-pitched murmuring we hobbyists call *flutter-baas* or *mama voice.*

## Second-Stage Labor

Second-stage labor begins when the cervix is fully open and a fluid-filled bubble appears at the ewe's vulva. This is the *chorion*, one of two separate sacs that enclose a developing fetus within its mother's womb (the other is the amnion). Either or both sacs might burst within your ewe or at the same time her lamb is born.

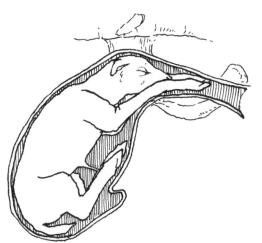

**A NORMAL DIVING-POSITION DELIVERY.**
*In a normal front-feet-first, diving position delivery, the tip of a hoof appears inside the chorion (or directly in the vulva if the chorion has already burst), followed almost immediately by another hoof and then the lamb's nose tucked close to his knees. Once the head and shoulders are delivered the rest of the lamb slips out.*

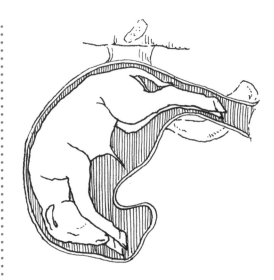

**A NORMAL HIND-FEET-FIRST DELIVERY.**
*In a normal hind-feet-first delivery, two feet followed by hocks appear. The lamb's umbilical cord is pressed against the rim of his mother's pelvis during this delivery, so gently help the lamb out once his hips appear. Otherwise, these are both textbook deliveries and rarely require assistance.*

Soon after the water bag appears, the ewe will lie down and roll onto her side as each contraction hits. She'll ride out the contraction, then rise and repeatedly reposition herself until she finds a position she likes. Once she does, she may roll up onto her sternum with her forelegs tucked beneath her between contractions but she'll usually remain lying down on her side, often with her head turned back toward one side. Some ewes deliver standing up or in a squatting position, and that's okay, too.

## When Things Go Wrong

Sometimes, however, things go wrong and you should always be mentally prepared to assist. Sheep aren't cows, and you can't pull a lamb with tackle the way you'd pull a calf. If you do pull, slather lots of lube inside the birth canal alongside the lamb (you can't use too much), and pull *only* while the ewe is pushing. Don't pull straight back, pull down in a gentle arc toward your ewe's hocks. To pull a lamb, grasp his legs, preferably above the pasterns but below the knees or hocks, and pull only during contractions. In an emergency you can also pull on his head.

Before examining your ewe internally, I can't stress often enough that you must make sure your fingernails are as short and smoothed as possible and you've removed your watch and rings. Swab her vulva using warm water and mild soap or a product such as Betadine Scrub. Slip on a medical exam glove or scrub your hand and forearm with whatever you used to clean the ewe.

Then, liberally slather the glove or your hand and arm with lube and place lube inside her birth canal, too. Pinch your fingers together, making your hand as slim and elongated as possible, and gently ease your hand into her vulva, advancing *only* between contractions.

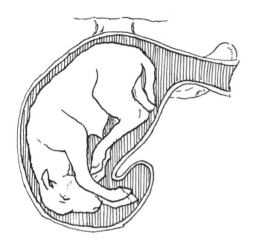

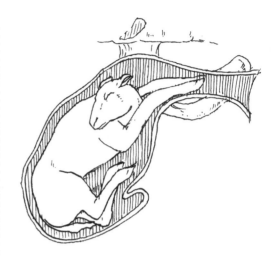

**TRUE BREECH PRESENTATION (BUTT FIRST, HIND LEGS TUCKED FORWARD).** *Some ewes can give birth in this position but most can't. It's difficult to reposition a true breech presentation, so call your vet. If you must reposition this lamb by yourself, try to elevate your ewe's hindquarters before you begin.*

*1. Push the lamb forward, then work your hand past his body and grasp one hock.*

*2. Raise the hock up and rotate it out away from his body.*

*3. While holding the leg in that position, use the little and ring fingers of the same hand to work the foot back and into normal position; repeat on the opposite side.*

*4. The umbilical cord will be pinched, so pull this lamb as quickly as you safely can.*

**HEAD BACK.** *To correct this problem, attach a lamb snare to the front legs at the pasterns (or tie a length of cord to each pastern) so you don't lose them, then push the lamb back as far as you can and bring his head around into position.*

*Sometimes the lamb's front legs are presenting properly but his head is bent forward and down. This is more difficult to correct. If you have to reposition this yourself, do so in the same manner, but it's a tough job, so enlist your veterinarian's help if you can.*

**ONE LEG BACK.** *Older ewes and large ewes delivering small lambs may deliver a lamb with one leg back. It's best to reposition the trailing leg for first-timers and small ewes delivering large lambs. To do this, push the lamb back, hook your thumb and forefingers around his knee, and use the rest of your fingers to draw the leg forward into place.*

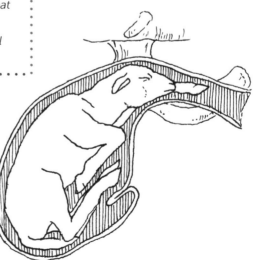

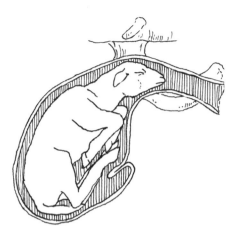

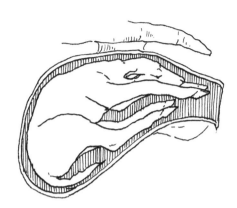

**FRONT LEGS BACK.** *Some ewes can deliver a normal-size lamb with one leg back, but if the ewe is young or the lamb is big, push the lamb back far enough to allow you to cup your hand around the trailing hoof and gently pull it forward.*

*It's harder when both front legs are back and only the head is in the birth canal. Elevate the ewe's hindquarters if you can, push the lamb back into her uterus, and bring the trailing legs forward one at a time.*

**ALL FOUR LEGS AT ONCE.** *Attach a snare to one set of legs, making certain you have two front legs or two rear legs, not one of each, and then push the lamb back as far as you can. Reposition him for either a diving position or hind-feet-first delivery, depending on which set of legs you have in the snare.*

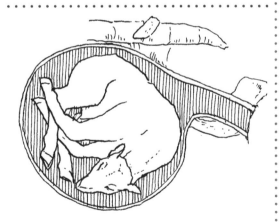

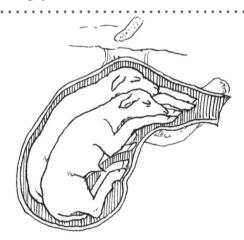

**CROSSWISE.** *This is definitely a job for your veterinarian. If he isn't available, push the lamb as far back inside the ewe as you can (elevating her hindquarters will help) and determine which end is closest to the birth canal, then begin manipulating that end into position. These lambs are usually easier to deliver hind feet first.*

**TWINS COMING OUT TOGETHER.** *Attach a snare to the front legs of one lamb (follow the legs back to make sure they're attached to the same lamb). Push the other lamb back as far as you can and bring the snared lamb into a normal birthing position.*

*If one lamb is reversed, follow the same protocol keeping in mind that it's usually easier to pull the reversed lamb first. If both are reversed, pull the closest lamb first.*

Figure out which parts of the lamb are present in the birth canal. If his toes point upward and the big joint above them bends away from the direction the toes are pointing, they're forelegs. If his toes point down and the major joint bends in the same direction, they're hind legs.

Follow each leg to the shoulder or groin to make sure the parts you're feeling belong to the same lamb. If they do and if you can manipulate him into a normal birthing position, do so. Then, if the ewe is exhausted gently pull the lamb with each contraction; otherwise, sit back and allow the ewe to deliver the lamb herself. However, if you have to pull one lamb and others are waiting to be born, call your vet.

When repositioning a lamb, cup your hand over sharp extremities such as hooves and work carefully and deliberately. Any time you have to assist inside a ewe, you must follow the birth with a course of antibiotics. Ask your vet for advice.

## Helping in an Emergency

Your ewe's first lamb should be born within an hour after hard labor commences. Some producers swear by the 30-30-30 rule:

- 30 minutes after hard labor begins for the water sac to appear
- 30 minutes for the first lamb to be born
- 30 minutes for second and subsequent lambs to be delivered

Some ewes, however, especially first-timers, take longer at each step.

*This farmer has his hands full, as he juggles a baby while feeding two lambs. The back of the photo reads, "Matilda Pauline, 4 months old."*

### If the Ewe Stops Straining

You have two options if your ewe's contractions stop before a properly positioned lamb already in the birth canal is born. The best is to call your veterinarian without delay. He'll give her a shot of a drug called *oxytocin* to start her contractions again.

If the veterinarian can't come, slather on plenty of lube and then go ahead and slowly pull the lamb. This is not easy and it's very dangerous for your ewe. If you or the lamb tears her internally, she will die. Attempt this only as a last resort.

**DISPORTIONATE SIZE.** Sometimes a lamb is too big or a ewe's pelvic opening is too small for a lamb to be born in the normal manner. If you suspect this is the case and the lamb is alive, your veterinarian can deliver the lamb by C-section (caesarean); if the lamb is dead, he'll dismember it and deliver it in sections, thus saving the ewe.

### Ringwomb

Sometimes a ewe's cervix fails to fully dilate. While some shepherds attempt to gently work it open with their fingers, don't. Call your veterinarian. If he can't dilate the ewe's cervix, he can deliver her lambs by C-section. Ringwomb should be left to the pros.

## Here Be Lambs!

When the first lamb arrives, pull the chorion away from his face (tear it open if the lamb is still fully enclosed when he pops out) and strip birthing fluids from his nose by firmly squeegeeing your fingers along both sides of his face from just below his eyes down to his nose. If he's struggling to breathe, use the bulb syringe from your lambing kit to remove excess fluid from his nostrils. If he's really struggling to breathe or not breathing at all, take a secure grip on his hind legs between his hocks and pasterns (he will be very slippery, so hold on tight), place your hand behind his neck to support it as best you can, and swing him in a wide arc to clear fluid from his airways and jumpstart his breathing.

A surprising number of lambs from assisted lambings appear to be dead; in fact their hearts are pumping, but they simply haven't started to breathe. Swinging usually fixes that. Tickling the inside of his nostrils with a piece of straw

**LEARN FROM THE PAST**

Often a lamb has a hard struggle at birth and arrives in this new world almost exhausted, lying without any signs of lung action. The shepherd has assisted the ewe in bringing the lamb forward, but it seems almost, but not quite dead. All that shows the lamb to be alive may be a single quiver. Now is the time he must act quickly to revive the lamb. The first thing is to clean all phlegm out of its mouth, then he must hold the mouth open with his two hands and blow gently three or four times into it to start up lung action. Now he must lay it on its belly and gently beat it slightly with his two hands, one on each side on its heart girth right back of the shoulder, and if it does not commence to breathe, he should blow into its mouth again. If there is the slightest bit of life left in the lamb, he will revive the lamb by this method. Many such lambs that at first sight appeared to be dead, have been revived by the writer in this method.

— Frank Kleinheinz, *Sheep Management: A Handbook for the Shepherd and Student* (1912)

or hay works, too. Don't give up. If you keep stimulating these lambs, chances are they'll start to breathe and be perfectly okay.

Once he's breathing, place the lamb in front of his mom so she can clean him. The taste and scent of her newborn creates a maternal bond. At some point she may leave him to deliver another lamb. This is normal. Simply place both lambs in front of her after the second one has arrived.

Once all lambs are born, think "snip, dip, strip, and sip." Snip the umbilical cord to a manageable length if it's overly long (about 1½ inches [about 4 cm] is just right), dip the cord in iodine (fill a shot glass or film canister with iodine, hold the container to the lamb's belly so the cord is completely submersed, then tip the lamb back to effect full coverage), strip your ewe's teats to make certain they aren't plugged and that she indeed has milk, then make sure all lambs sip their first meal of colostrum within an hour or so after they're born.

The ewe's placenta should pass within an hour or two after her final lamb arrives. Most ewes eat the afterbirth if you let them but this a choking hazard, so try to remove it before she gets to it. (I once had to give the sheep-y version of the Heimlich maneuver when a ewe tried to gobble her newly shed placenta and nearly died.) Wear gloves or pick it up with a plastic bag. Burn or bury it; don't let your cats or dogs eat it. If the placenta hasn't been delivered within 12 hours (you'll know because bits will still be dangling from her vulva), call your veterinarian.

## Post-Lambing

Every shepherd must be aware of two post-lambing infections: metritis and mastitis (see page 102). They don't occur with every pregnancy but are serious conditions when they do occur, so be prepared.

---

### A Post-Dystocia Pick-me-up for Ewes

A British friend suggests the following elixir for ewes stressed by a difficult delivery. She uses it all the time — and says it works for frazzled shepherds, too!

2 aspirins crushed and dissolved in
½ cup coffee

2 tablespoons honey

2 tablespoons whiskey

Stir well and dribble it into the ewe's mouth using a small syringe with the needle removed.

## Happy Mother's Day

**Hope had been acting "off"** for a few days, staring introspectively into space and occasionally taking herself away from the others as though she might lamb, so I watched her closely. On Mother's Day morning she went into labor. I brought out my lambing kit and camera and settled in for another routine birth. It was not to be.

Hope got down to business, and her water bag burst. From that point there should have been front toes showing within 30 minutes. There weren't. When I examined her internally, I found nothing but an enormous head, aligned sideways instead of vertically with Hope's tail as it ought to be. I called John to come help and the ordeal began.

I pushed the lamb back into Hope's uterus and began feeling for feet. There was only one. I worked it into position and kept searching. Nothing. It was desperately hot, Hope was screaming, and there was no second leg. Because she was a first-timer and tight, there was no way she could deliver the lamb with one leg back, nor could we pull it.

After an excruciating amount of time, John found the second leg and worked it into position, but there was too much lamb to fit through the birth canal. If we couldn't pull the lamb, Hope would die. I promised Hope that if she survived this lambing, I would never, ever breed her again.

We hauled out another bottle of Super-Lube, I held Hope, and John slowly, painfully, managed to pull the lamb, which flopped onto the straw, looking very, very big — and dead. I quickly stripped fluid from his nostrils and checked: he had a heartbeat but wasn't breathing. We swung him, we shook him, and then we tried a primitive CPR technique I had learned from a century-old book (see Learn from the Past, page 135).

It worked! The lamb gasped, sneezed, and shook his head. We set him in front of his exhausted mom, and as if nothing was wrong, she began talking to the lamb and cleaning him.

We named the little guy Arthur, short for Wolf Moon Wee Mad Arthur (for a Nac Mac Feegle character in Terry Pratchett's Discworld books). He was strong once he began breathing and was up on his feet in record time.

There was blood, more blood than a normal birth, so I was very much afraid Hope would die. Yet she struggled to her feet in less than half an hour, and soon she was feeding baby Arthur. We treated her with penicillin and Banamine and held our breath. After 5 days I began breathing easier. Now I know what assistance "à la Herriot" is really like. And it was a Mother's Day I (and Hope) will always remember.

# Caring for Lambs

# 10

*To rise with the lark, and go to bed with the lamb.*

— Nicholas Breton, *Court and County* (1618)

**M**ost lambs stand within 10 to 30 minutes after birth and immediately begin seeking a teat. Encourage this; lambs should ingest colostrum within the first hour or so after birth.

Neonatal lambs are wired to seek food in dark places. Help weak or disoriented lambs by holding them near the ewe's udder, but let them find the teat by themselves (most resist a teat placed directly in their mouths).

Lambs bunt their dam's udder to facilitate milk letdown. A rapidly wagging tail generally means a lamb is suckling milk. After feeding, contented lambs snooze. A lamb who cries, noisily suckles a great deal of time, or constantly worries his dam's udder isn't getting enough to eat. This lamb will require supplementary feeding with a bottle and formula until he's old enough to eat sufficient solid food to fill the gap (we'll talk about bottle lambs later in this chapter). If you aren't sure a neonatal lamb is getting enough to eat, pick him up an hour or so after he begins attempting to nurse; he should have a full, round tummy.

## Dealing with Rejected Lambs

Sometimes ewes refuse to mother their lambs. Just as Frank Kleinheinz says, this sometimes occurs when a firstborn wanders off while his mom is giving birth to or cleaning a second lamb. Other causes include:

The ewe may have a painfully colostrum-engorged udder and it hurts when her lamb tries to nurse. In this case, partially hand-milking both sides of her udder (saving the colostrum to freeze) may set things right.

Some lambs are born with sharp-edged teeth that, again, cause pain when she tries to nurse. Check the lamb's teeth and use an emery board to smooth sharp edges.

First-time moms are sometimes afraid of their lambs. Others are so entranced by their new additions that they continually maneuver to keep the lamb in front where they can see him. Halter or collar and secure these ewes, reassure them, and help their lambs find a teat. Once her lamb nurses, most first-time moms get the drift.

Ewes with twins or triplets, when left with other sheep, often disown one of their lambs. In the majority of instances the stronger lamb comes first, and soon after birth it looks for its first meal. Its mother, however, is in pains to deliver another lamb, and therefore she will not move away from the nest which she has selected for lambing, which is generally in one corner of the barn. The mother does not follow her new-born lamb, but the other inquisitive sheep flock around to see the newcomer and often lead it away. The new-born lamb thus loses track of its mother, and the mother likewise loses the smell of her lamb and refuses to own it when she meets it again, since ewes recognize their lambs only by their smell and voice. Ewes should, therefore, be put away separately either in lambing pens or in a special place temporarily prepared for them by means of hurdles placed in corners of the barn. Here they are kept for a couple of days until mother and lambs are thoroughly familiar with each other.

— Frank Kleinheinz, *Sheep Management: a Handbook for the Shepherd and Student* (1912)

A ewe who has had a long or painful delivery may not be very interested in her lamb. As long as she isn't aggressive toward him, milk her, bottle- or tube-feed colostrum to the lamb, and leave him with his mom, as she'll probably mother him when she recovers.

Sometimes ewes with multiple lambs will accept one or even two lambs and reject the rest — possibly because she instinctively knows she can't support so many lambs.

Some ewes, especially young or flighty ones, reject lambs if you interfere with their bonding process. If you don't know how a ewe will respond, clear mucous from her lambs' faces and dip their umbilical cords, then get out and leave the lambs alone. Altering a lamb's appearance also upsets some ewes, so if you fit lambs with lamb coats, stand by to make sure the dam accepts them.

## Finding a Foster Mom

Keep an eye on a ewe who seems antagonistic toward a lamb. She may start by gently nudging him aside but progress to flinging him against the sides of her lambing jug. If a lamb seems in danger, remove him before she hurts him and bottle-feed him or graft him onto another ewe.

A ewe who has recently lost her own lamb may accept an orphan or rejected lamb as her own. Old-timers used to skin the dead lamb and tie his skin over the graftee, a messy and often unsuccessful maneuver. A better ploy is to secure the ewe in a head gate or halter and tie her so she can't get away from the lamb or butt him and then allow him to nurse his fill. This must be repeated at every feeding until the ewe accepts the lamb, which may occur a day or two later, after he's ingested enough milk that she detects her scent in his feces. Still, it's far from foolproof. You may have to bottle-feed this lamb (see page 144).

## Lamb Poop 101

Within a few hours of birth, newborn lambs pass a black, tarry substance called *meconium*. Watch for it. If a lamb seems to be straining without success, you need to give the little guy a hand. The best thing we've found is a smidge of commercial human infant enema product inserted in the rectum. (We buy Fleet enemas packaged for human infants. They're designed for insertion, though a typical lamb doesn't need a full human infant dose.) Set the lamb down; if he doesn't pass black goo within five minutes, try again until he does.

Beginning a day or so after birth and continuing for several days, lambs pass thick, yellowish, pudding-consistency poo. Some

### Experienced Ewe's Weird Reaction

**Wolf Moon Wren gave birth** to twin black ram lambs today. It's going to be cold tonight so I put lamb coats on Wren's boys. Then, since I had an extra coat, I decided to put it on Fosco, Shebaa's white lamb born the day before yesterday. Keep in mind that Shebaa is 7 years old and this is her sixth lambing. And she's normally a very sensible ewe.

I scooped up the lamb, dressed him in a dark purple lamb coat, and set him down. Shebaa rose up on her tiptoes and her eyes bulged out on stems. She began backing up s-l-o-w-l-y as Fosco toddled toward her, then she whirled and bolted. Fosco dashed after her, causing her to run faster — and faster and faster.

Old Rebaa (who is Shebaa's and Wren's mom) hove to her feet and hobbled after them. Shebaa doubled back past the shelter and as mom and trailing lamb streaked past, I scooped up the lamb. Shebaa careened around the top of the paddock again, then screeched to a halt and whirled ("Where is my lamb!"). Rebaa rammed her like a freight train. She head-butted Shebaa several times ("What is *wrong* with you? I raised you better than this!") while I divested Fosco of his jacket.

When I set Fosco down, Shebaa rushed to collect him, then whisked him into the loafing shed while Rebaa and I watched in amazement (me) and annoyance (Rebaa).

And the weird thing is that I know I put a stretchy blue, navy, and white striped dog sweater on Baabara, Shebaa's first lamb, because I have a picture of newborn Baabara wearing it. But Baabara was black. I guess the dark purple coat changed Fosco into the bogeyman because he's white and she noticed the difference much more dramatically.

I could see a first-time mom doing this, but Shebaa? Not in a million years! So Fosco is on his own. At least he's wearing his own wool coat.

ewes meticulously clean their lambs; if yours do, be very, very grateful. Otherwise, gooey lamb poo gets caught in a lamb's tail, where it dries, sometimes gluing the tail to the lamb's hindquarters and in any case, blocking his anus (and in ewe lambs, her vulva), making it impossible for the lamb to eliminate.

Prevention is better than cure. Greasing the bottom of a day-old lamb's tail and the area around and below his anus with petroleum jelly helps keep pudding-poop from accumulating — sometimes. If it doesn't, and you don't manually clean the lamb, he could die. It's a messy job, but it needs to be done.

Clamp the lamb under one arm and use plain warm water and a washcloth to clean him. For dried poo, soak with a warm, wet washcloth, and then carefully pick off remaining residue. Be careful, it's easy to tear lambs' delicate skin.

Some ewes give so much milk, especially after giving birth to a single lamb, that their lambs scour (have diarrhea) badly. Don't feed such ewes grain or rich, legume hay; switch them to a plain, grass hay diet for several days to a week until the lamb catches up with his mother's milk production.

In the meantime, treat the lamb with doses of Pepto-Bismol appropriate to his size. (We give our small-breed lambs 1 cc using a 3 cc syringe with the needle removed.) If necessary, repeat the initial dose at hourly intervals until scouring subsides. Resist the temptation to give a lot the first time, lest you constipate the lamb.

## Tails It Is

Wild sheep such as Mouflons and Rocky Mountain Sheep have short, hair-covered tails;

domestic hair sheep have them, too. European short-tailed breeds have short, fluke-shaped tails with little or no wool covering. However, centuries of selection for wool production led to most sheep having longer, thicker, woollier tails that become saturated with urine because extra-woolly tails are too heavy to raise out of the way, or they become encrusted with manure if the sheep has diarrhea or is housed in wet, dirty conditions.

Enter the blowfly, a particularly nasty external parasite that lays its eggs in wounds and damp fleece, causing a horrible condition called *fly strike* (see chapter 7). This is why most shepherds dock (shorten) their lambs' tails: to prevent fly strike. It isn't done strictly for "looks."

Dock tails one or two days after birth; the older the lamb, the more painful the operation. Some shepherds simply snip the tail off

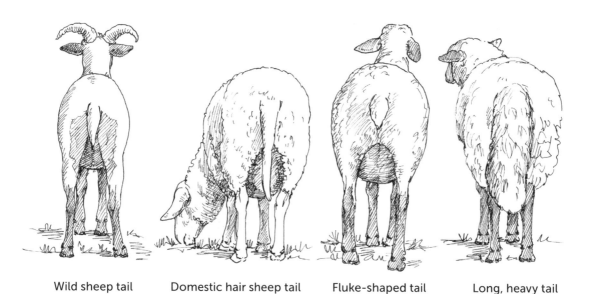

Wild sheep tail     Domestic hair sheep tail     Fluke-shaped tail     Long, heavy tail

and treat the stump with blood-stop powder. We prefer banding, which is the application of a thick rubber band using a tool called an *elastrator*. This cuts off circulation to the tail and in about two weeks the tail falls off. Newly

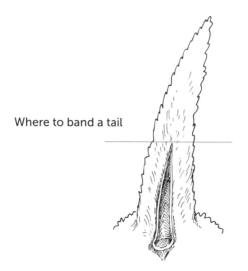

Where to band a tail

banded lambs nearly always take banding in stride; most don't seem to feel any pain, while a few do. In either case, the tail becomes numb and pain-free in a few minutes. Lambs whose dams were boosted with CD/T toxoid a few weeks prior to lambing need no further protection. Barring that, it's wise to give lambs an injection of tetanus antitoxin at banding time.

At one time show sheep were docked so short that virtually no tail was left; however, ultrashort docking affects the muscles and nerves of the anus and prevents the sheep from lifting her tail to direct fecal pellets away from her hindquarters. Additionally, short docking is said to contribute to rectal and vaginal prolapses. Tails should be left at least long enough to cover a ewe's vulva and a ram's or wether's anus.

The American Veterinary Medical Association, American Association of Small Ruminant Practitioners, and American Sheep Industry Association agree that tails should be removed no shorter than the end of the tail fold that is farthest from the lamb's hindquarters. To find it, turn the tail over and look for where the bare patch of skin under the tail ends in a point. We find that's still a little short once the lamb is full-grown, so now we band at least an inch below where the patch of bare skin ends.

## Feeding Dam-Raised Lambs

Some people recommend creep feeding grain to lambs. A *creep* is a feeding area designed with barriers around it so that only lambs can enter and eat whenever they please. We don't creep feed our lambs because our little ewes are excellent milkers. If you choose to creep feed, remove uneaten grain and replace it with clean feed once a day. And watch those ewes; some are very adept at invading supposedly ewe-proof creep feeders.

Because our Ozark pasture is not the best, we do, however, supplement lactating ewes' diets with dehydrated alfalfa and a small amount of grain. Lambs quickly learn to sample their mothers' feed, so they're familiar with concentrates by the time they're weaned.

Once weaned, we feed lambs all the top-quality Bermuda grass hay they'll eat, with a measured amount of grain and dehydrated alfalfa twice a day until they're a full year old. Again, discuss feeding strategies with your County Extension agent to devise a feeding plan based on available feedstuff.

## Weaning Dam-Raised Lambs

We wean our lambs by dividing them by gender (ram lambs and wethers can bunk together) and putting them in separate paddocks that abut the ewe paddock. Their mothers visit them along the fence line, but most ewes lose interest within a few days.

Ram lambs of some breeds are sexually precocious and should be weaned by about 12 weeks of age. Ewe lambs can be older. Some breeders leave ewe lambs with their dams until the ewes wean their daughters by no longer allowing them to suckle; this usually happens between 4 and 6 months of age.

### DID EWE KNOW?

#### Superfetation

*Superfetation* is the simultaneous occurrence of more than one stage of developing embryo in the same mammal; applied to sheep, this means a ewe is carrying lambs conceived at two different times.

If this happens, a ewe has probably conceived on two successive heat cycles, and she will give birth to another lamb (or lambs) 13 to 19 days after her first lambs arrive. Some ewes accept their lambs from both lambings, though usually they're disinterested in one or the other.

So, if you find a lone newborn or two crying for mama, check the backsides of all your ewes, not just the ones who haven't yet lambed. Although an uncommon occurrence, it happens often enough that you should be aware.

## Raising Bottle Lambs

Bottle lambs might be called *bummers* or *poddys*, depending on where you live. Large-scale producers often give orphans and lambs from multiple births to family farm shepherds like you and me or will sell them at bargain basement prices. This is a great way to build a flock of tame, people-oriented sheep, but even with free lambs, it isn't cheap and it certainly isn't easy, as a bottle lamb will depend entirely on you for 10 to 12 weeks (or more, if you choose).

For the first few weeks you'll be warming formula and holding bottles four to six times every 24 hours, and some of those feedings will fall in the dead of the night. A bottle lamb will crave your attention. If you can't spend a lot of time with him, consider adopting two lambs or a lamb and a goat kid (the kid may push the lamb around a bit, but they'll still become best friends) so they'll have each other for company when you're not around. And no method of feeding bottle lambs is inexpensive.

### The Importance of Colostrum

All newborn lambs, dam- or bottle-raised, need colostrum, the thick, yellow, nutrient- and antibody-rich milk that ewes produce for the first few days of each lactation. Because lambs are born with no disease resistance of their own and because they absorb antibodies from colostrum for only 12 to 36 hours after birth, every lamb should ingest 10 to 15 percent of his body weight in colostrum during that time frame; preferably at least half of that amount within 4 to 8 hours of birth.

Responsible shepherds make certain that the lambs they sell as bottle babies receive enough colostrum before they leave the farm, but be sure to ask. If your lamb didn't get enough colostrum, he'll be at risk for disease until his own immune system kicks in around 7 to 8 weeks of age.

To err on the side of caution, collect colostrum from a newly lambed ewe or visit a goat or cattle dairy farm and buy at least 16 ounces

### DID EWE KNOW?

## Lambing Lore

According to Icelandic folklore, if the earth remained buried in snow all winter, the new season's lambs would be white. Open winters spawned colored lambs, and patchy snow cover, spotted ones.

In western European folklore, if the first lamb of the season was first seen facing the viewer, it was a good omen; facing away, bad luck. If the first birth was twins, so much the better (unless one or both were black).

In British folklore, black sheep were said to bring good fortune, whereas in most other European countries the opposite was true.

Ewes and their milk were associated with the Celtic goddess Brigit and her spring festival of Imbolc (February 2), which means "ewe's milk."

Shepherds in Europe and the United Kingdom were sometimes buried with a tuft of wool in their hands, signifying their devotion to their charges and to show they were shepherds, thus excusing occasional lapses in church attendance because they couldn't leave their flocks during lambing.

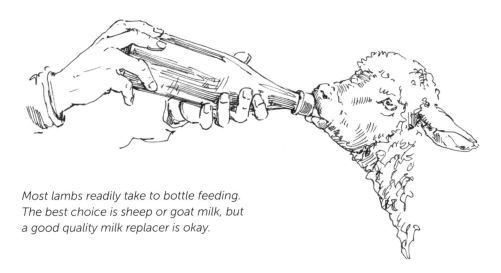

*Most lambs readily take to bottle feeding. The best choice is sheep or goat milk, but a good quality milk replacer is okay.*

(0.5 l) of colostrum to store in your freezer. Bag and freeze it as soon as you can. Individual 2- or 3-ounce feedings stored in resealable sandwich bags work well. Squeeze out excess air, seal them, and store individual baggies in a larger zip-lock freezer bag. Store the larger bag in a non–frost-free freezer (the constant thaw and freeze cycles in a frost-free freezer compromise colostrum integrity). Properly stored, the antibodies in frozen colostrum remain viable for at least a year. Thaw each feeding at room temperature or by immersing the baggie in warm water; don't heat it over the stove or in the microwave.

Farm stores and livestock supply catalogs carry colostrum boosters and replacers based on freeze-dried, whey-based, or serum-derived cow colostrum. They aren't the best approach but if you use one, Colorado State University research suggests that bovine serum-based supplements such as Colostrix, Immustart, and Lifeline work reasonably well.

An even better bet is Goat Serum Concentrate, an injectable immunotherapy product manufactured by Sera, Inc. It's packaged for goats but since goats and sheep share a host of diseases, it works for sheep as well. Injected subcutaneously (not into muscle) as soon after birth as possible, it protects lambs from a broad spectrum of viral and bacterial agents.

### Choosing Milk or Formula

The best food for lambs is fresh or frozen sheep or goat milk. Or use pasteurized, homogenized, whole-fat cow milk. Some folks feed it straight from the jug, but it's better to boost its fat content by adding one part dairy half-and-half to four parts whole milk. Lambs thrive on this combination and it's a snap to store and prepare.

Milk replacers formulated for lambs work, too; however, choose a good one. Inexpensive milk replacers are usually based on soy rather than milk protein; lambs can't digest it and they'll fail to thrive. Better products like Advance Lamb Milk Replacer (we use this when we don't have fresh or frozen goat milk at hand), Purina, Land O'Lakes, and Merrick's Super Lamb are excellent choices. Despite what their labels claim, *never* use one-kind-fits-all milk replacers or products formulated for the young of other species. Besides not being formulated for a lamb's nutritional needs, these contain dangerously high amounts of copper. Always choose milk protein–based products labeled specifically for sheep.

Abruptly switching from milk to replacer, or from one brand of replacer to another, upsets lambs' tummies. The result is diarrhea, also known as *scour*. It's not pleasant and more important, it's dangerous. Diarrhea leads to dehydration and serious problems such as enterotoxemia; if you must switch, do it gradually over the course of several days. Ideally you should stick to the same milk or formula as long as you bottle-feed your lamb.

### Bottle-Feeding Equipment and Technique

Our favorite lamb-feeding device is the Pritchard flutter-valve nipple. This funny-looking, soft, red rubber nipple works well for weak or newborn lambs yet it's fine for older lambs, too. The Pritchard teat's yellow plastic base can be screwed onto any glass or plastic bottle that has a 28-mm (approximately 1 inch)

opening — plastic soda pop bottles work well. A metal ball bearing inset in a hole in its base neatly regulates air and milk flow so that lambs don't suck excess air or choke.

It's also fine to go with what you know. Bottles and nipples designed for human infants work for lambs of small- to medium-size breeds. When your lamb chews up an everyday infant-nurser nipple, it's convenient that you can drive to any grocery or corner pharmacy store and pick up more.

What you don't want is the type of hard, black rubber "lamb nipple" available through livestock catalogs and farm stores. Small lambs can't suck vigorously enough to nurse from them, while older ones jerk them off of feeding bottles and choke.

To prepare a Pritchard nipple for a newborn, carefully snip off the tiniest bit of the tip leaving just enough of a hole for him to nurse; with other nipples, slightly enlarge the hole using a hot needle.

Buy a bottle brush to scrub the bottles, a measuring cup for mixing formula, a funnel, and chlorine bleach to keep everything squeaky clean.

*Once a lamb figures out how to suckle from a bottle, he'll eagerly empty it at every feeding.*

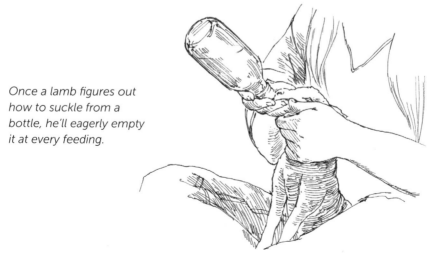

To bottle train a lamb, sit cross-legged on the floor with the lamb tucked between your legs facing away from you, his front legs straight and his butt on the floor. Cup your left hand under his jaw and open his mouth using your left thumb and forefinger. Insert the nipple with your right hand, then balance and steady it with the fingers of your left hand with the left palm still under his jaw. This way he's less likely to spit out or otherwise lose the nipple.

Encourage the lamb to lift and tip back his head. Lambs function as single-stomached animals rather than ruminants. When his head is up and back, a band of muscle tissue called the *esophageal groove* closes and, bypassing his undeveloped rumen, funnels milk straight from his mouth to his abomasum (the only stomach compartment containing digestive enzymes).

Don't let milk pour into his mouth, however. If he can't swallow fast enough, it might spill into his lungs. Hold the bottle as level as possible while still keeping fluid in the bottle cap and nipple.

If your lamb won't nurse, placing a towel or an assistant's hand over his face may help him get started (this simulates the darkness of a ewe's underbelly). Or dab colostrum or formula on his lips; give him a taste but don't shoot a stream in his mouth. Older lambs that have nursed their moms are notoriously hard to bottle train. Persevere. If an older lamb refuses to nurse and he's otherwise healthy, place him back in his quarters and wait an hour or two before trying again; if he's weak or sick, tube-feed him, as described on the next page.

## Cleanliness Is Key

Improper bottle-feeding quickly leads to diarrhea, so preparing meals with care is important. Measure ingredients for homemade formulas and mix lamb milk replacers according to package directions.

Keep nipples, bottles, and mixing utensils clean — very clean — by washing them in hot, soapy water after every feeding. At least once a day immerse bottles and utensils in 10 parts water to 1 part chlorine bleach solution. Don't bleach nipples; bleach solution degrades them.

### Watch Those Nipples!

Test the nipples, especially soft rubber Pritchard teats, before every feeding. Affix the nipple to a feeding bottle and then tug on the rubber part to make sure it doesn't tear apart. When nipples fail, which they do after weeks of hard use, they usually do it all at once, while a lamb is lustily feeding. Then he swallows the nipple. It's happened to me twice, though without ill effect. Don't trust those nipples!

Pritchard teat

## Tube-Feeding Equipment and Technique

It's best to have tube-feeding apparatus on hand because a weak lamb who can't suck must be tube-fed, a process that sounds much scarier than it is. You'll need an empty 60 cc catheter-tip syringe (60 cc = 2 fluid oz, the correct dose per feeding for most newborn lambs) and a soft plastic feeding tube. Just the syringe (your vet should have them) and 28 inches (71 cm) of soft, ¼-inch (6 mm) plastic tubing from the hardware store will suffice in a pinch.

Here's how you do it (recruit a helper if you can).

1. Place a premeasured amount of colostrum in a glass measuring cup and remove the plunger from the syringe.
2. Lay the tube along the lamb's side and measure from just in front of his last rib to his mouth. Mark that point with tape or a marker.

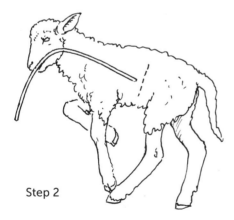

Step 2

3. Place the lamb in your lap on either his left or right side (he can sit up if he's strong enough) facing away from you. See illustration for how to hold his head.

4. Gently insert your thumb between his lower teeth and upper dental palate to open his mouth.
5. Keeping a finger in his mouth, carefully insert the tube via the side of his mouth and feed it down his throat very slowly, allowing him to swallow it if he can. Keep passing the tube until the mark you made is even with his mouth.

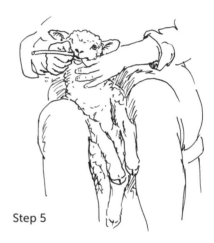

Step 5

6. With the tube in place and while supporting the lamb as best you can (this is where a helper comes in handy), attach the barrel of the syringe and add the colostrum.

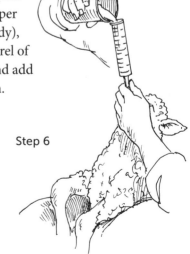

Step 6

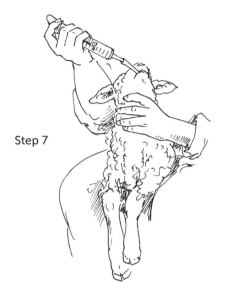

Step 7

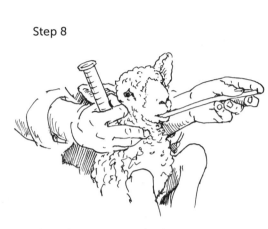

Step 8

**7.** Hold the syringe up high to allow the fluid to gravity-feed into the lamb's stomach.

**8.** When the syringe and tube are empty, pinch the tube and draw it back out. Don't forget to pinch; you don't want to spill fluid into his lungs.

## Be Careful with the Tube

You mustn't insert the tube into his trachea (windpipe), as fluid in his lungs can lead to pneumonia and possibly kill him. (This is also why syringing colostrum into his mouth is a bad idea.) A lamb's esophagus is soft; keeping your fingers on his throat, you should be able to feel the tube descending and know that everything is okay.

If the tube does inadvertently enter his trachea, the lamb will struggle violently and the tube will halt midway to your mark. Gently pull it out and start again.

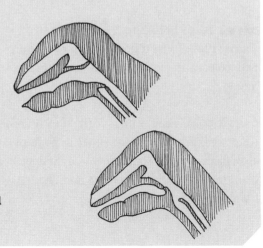

## Preparing Meals

Whether bottle- or tube-feeding, measure the amounts you feed your lamb, and feed at evenly spaced intervals. Lambs require 15 to 20 percent of their body weight in milk or formula daily. This schedule works for a small- to medium-size breed.

**DAYS 1 TO 2:** Feed 2 to 3 oz (0.06 to 0.09 l) warm (102°F [38.9°C]) colostrum or formula with colostrum replacer added, six times in 24 hours.

**DAYS 3 TO 4:** Feed 3 to 5 oz (0.09 to 0.15 l) warm milk or formula, six times in 24 hours; also, place a small, spill-resistant container of drinking water in his quarters and change it frequently to keep it fresh.

**DAYS 5 TO 14:** Feed 4 to 6 oz (0.1 to 0.2 l), four times in 24 hours, gradually switching to milk or formula straight from the refrigerator.

**DAYS 15 TO 21:** Feed 6 to 8 oz (0.2 to 0.24 l), four times in 24 hours; introduce green pasture or free-choice leafy alfalfa or high-quality grass hay and small amounts of high-protein (16–20%) dry feed.

**DAYS 22 TO 35:** Gradually work up to 16 oz (0.5 l) of milk or formula fed three times in 24 hours; continue until the lamb is 6 weeks old.

**WEEKS 6 TO 8:** If lamb is eating sufficient amounts of forage (grass or hay) and chewing his cud, at 6 weeks begin decreasing the amount of liquid offered at two feedings, eliminating them altogether by the end of week eight; leave the remaining feeding at 16 oz.

**WEEKS 8 TO 12:** Continue feeding 16 oz until the end of week 11; in the next seven days, gradually eliminate the last feeding.

## Living Accommodations

Consider keeping your lamb in your home for 4 to 6 weeks. Housed in a wire dog crate with a flip-open top, a folding canine exercise pen, or an old-style wooden playpen, and kept clean and dry, a lamb has surprisingly little odor.

A lamb kept indoors bonds with humans, learns to interact with other household pets, and is just down the hall when his 3 AM bottle is due. That counts for a lot when your barn is a long trudge from the house.

Bed your lamb's indoor quarters with towels or pieces of old blankets. During the first week or so, when he's producing yellowish, pudding-consistency poo, you'll have to change his blankets fairly often. As he begins nibbling grass or hay he'll begin ejecting pellets you can scoop up with a tissue or paper towel, then you'll change bedding only as it becomes damp.

He'll need at least two tip-proof dog bowls or crate cups: one for water and one (or more) for dry feed.

You'll probably want to allow your (supervised) lamb to run loose in the house, but if puddles are a turnoff, use diapers during "indoor outings." Ewe lambs are easy: cotton or disposable diapers for comparably sized human infants do the trick (be sure to snip out a hole for baby's tail). A ram lamb can be fitted with a reusable, around-the-middle band designed for incontinent older dogs (most pet stores carry them).

About dogs: most are good with lambs as long as you're watching. They are, however, with the exception of trained livestock guardian dogs, wired to chase things that flee. Running play can be misinterpreted by the gentlest dog, so don't leave Fido and a lamb alone, unsupervised, in or outside of your home.

Eventually you'll move your lamb to your barn. Make certain his outdoor quarters are warm, safe, well ventilated, and free of drafts. If it's cold outside, don't use a heat lamp; they can and often do cause fires. Instead, fit him with a lamb coat.

Cardigan-style human baby sweaters make fine lamb covers; place your lamb's legs through the sleeves and button the sweater up his back. Dog jackets and woolen sweaters work for ewe lambs but usually aren't cut to accommodate a ram or wether lamb's penis. Jackets specifically tailored for lambs and kids are nice; some makers are listed in the Resources (page 196).

Wearing a lamb jacket

## Love Those Bottle Babies

**We brought baby Mopple** home today! Mopple is a Dorper-Katahdin ram lamb. I'm going to raise him to pull a wagon. Oddly enough, he shares a birthday (July 7) with Louie, my last bottle lamb, who is the sweetest guy you could ever imagine.

Louie's story is semi-unique. That is, it's unique among most flocks but not mine. His mama, Rebaa, is one of my very best ewes. She is affectionate, she grows lovely fleece, she's a great mom with plenty of milk, and her lambs are awesome. She does have one quirk: she gives me a bottle baby from each lambing.

Now most people will say I'm anthropomorphizing, but I honestly believe Rebaa knows how much I love bottle lambs, so she shares her bounty with me. Here's how it works.

Rebaa gives birth to twins or triplets every year. She happily cleans them up, feeds them a round of colostrum, then begins critically looking them over. Finally she selects a ram lamb and gently pushes him away. The first year I didn't understand, so I put him back under his mom. Rebaa hooked him with her face and gently flipped him away again. I gave her a cross word or two and returned him to the milk bar. She looked at me incredulously, whipped around, scooped him up, and slung him against the wall. "There," she as good as said, "that one is yours!"

Nowadays as soon as Rebaa makes her choice, I take that lamb and bring him to the house to bottle raise. Rebaa's happy, I'm happy, and we avoid a lot of unpleasantness. She's given me Dimitri, Baalki, and Louie in this manner. All grew up to be well-adjusted, friendly, and productive fleece wethers that I had the pleasure of raising as bottle babies in the house.

And Rebaa's daughters? They don't share their lambs with me. This shows that it pays to know your sheep's peculiarities and to observe them closely at and after lambing. Sheep can surprise you, even old pros like Rebaa.

# Part 4

# Enjoying Your Sheep

# Producing Quality Fleece 11

*And was that really — was it really a SHEEP that was sitting on the other side of the counter? Rub as she could, she could make nothing more of it: she was in a little dark shop, leaning with her elbows on the counter, and opposite to her was an old Sheep, sitting in an arm-chair knitting, and every now and then leaving off to look at her through a great pair of spectacles.*

— Lewis Carroll, *Through the Looking Glass* (1871)

If you keep wool sheep, give some thought to the extra steps it takes to produce high-quality fleeces to sell, to give to friends, or to keep for your own crafting needs. There's a good market for handspinner-quality fleeces and lots of things you can do with your own fleece, such as exhibiting it at fairs and fiber shows; spinning your own yarn with a drop spindle or spinning wheel; and using that homespun yarn to knit, crochet, or weave things of beauty. But they all start with a quality fleece.

## First Things First

Only healthy, well-fed, dewormed sheep who are free of external parasites grow great fleece. It's impossible to cut corners and produce quality fleece; it simply can't be done.

Consider fleece breaks. These occur when something traumatic happens to a sheep, like running a fever, being chased by dogs, giving birth to lambs, or experiencing an abrupt change in diet. Trauma causes each growing strand of fiber to become thinner and weaker where it emerges from the follicle. This thin spot stays in the strand as the fleece grows longer. Later, when a spinner tries to work with this fleece, it will break off uniformly at that point.

Such a fleece is useless to handspinners, so it's important not to stress sheep and to time procedures to minimize fleece breaks. For instance, change feed gradually over a period of a week or so instead of overnight, and shear ewes a month before lambing so the break occurs close to their skin. Working with your sheep to make them tamer cuts back on handling-related stress as well; relaxed, happy sheep grow stronger fleeces.

Buyers can be found for almost every kind of quality fleece, but if you're producing wool for yourself or to sell at specific venues such as to carpet weavers or fine-wool crafters, choose your breed and individuals within your breed accordingly. I've always wanted to learn to felt. Imagine my dismay when I learned that Cheviots and Scottish Blackface fleeces are two of the worst breeds of sheep for that process. I should have started with fine wool sheep such

as Merinos or, better yet, Karakuls. It's okay, I still love my sheep, but since I don't spin, knit, or crochet, I give their beautiful fleeces to handspinners who do.

Why don't I sell my fleeces? Frankly, I don't maintain my sheep under conditions conducive to growing the best handspinners' fleeces. My rough and rugged sheep traverse our steep, brushy Ozark pastures picking up bits of twigs and seeds in their coats — vegetable matter or VM in handspinners' parlance. Added to that, we don't coat our sheep (we'll talk about coats in just a moment), so they spoil their fleeces when they yank hay out of feeders and drop it onto each other's heads and backs. At shearing time our fleeces are in decent shape but not in the pristine condition that handspinners expect when buying fleece. Not every nice fleece is marketable spinners' fleece.

## Sheep Coats

Shepherds who expect to market fleeces at $8 to $20 a pound generally coat their sheep. These are not the canvas covers or slinky nylon

A sheep in sheep's clothing

# Shrek, the Wooliest Sheep Ever

No chapter about fiber would be complete without mentioning Shrek, the hermit Merino wether who grew a 15-inch (38 cm) staple, 60-pound (27 kg) fleece.

Shrek was born in 1994 at Bendigo Station, a sheep station near Terras, New Zealand. He left his flock (an act almost unheard of in close-flocking Merinos) and hid out in caves for six years. His solitary hiding place was discovered by shepherd Ann Scanlan in April 2004 and Shrek was returned to Bendigo Station. He became an international celebrity when he was sheared on national television so that his fleece could be auctioned, with the proceeds earmarked for a children's charity, Cure Kids.

A gentle guy, Shrek made many public appearances during the rest of his long life, including meeting Prime Minister Helen Clark. To celebrate his 10th birthday, 30 months after his initial shearing, Shrek was shorn again on an iceberg floating off the coast of Dunedin, New Zealand.

Shrek was the subject of three books and he figured prominently in a fourth. *Shrek: The Famous Hermit Sheep of Terras* was written by students at the local Terras school, where book sales still fund a second full-time teacher. Cure Kids continues to receive the royalties from two books penned by John Perriam of Bendigo Station, *Dust to Gold* and *Shrek: The Story of a Kiwi Icon*; in all, Shrek has raised more than $150,000 for Cure Kids to date.

Shrek, who suffered from arthritis as he grew older, was humanely euthanized on June 6, 2011, at 16 years of age. His stuffed remains are on display at Te Papa Tongarewa, New Zealand's national museum.

body sleeves you may have seen on club lambs at fairs and lamb shows, nor are they designed to keep sheep warm. Sheep coats are made of breathable, lightweight, durable synthetic fabric. They're designed to cover the prime parts of a sheep's fleece, leaving only the head, legs, belly, and a little bit of butt sticking out. To fit a sheep with a cover, pull the cover's front opening over the animal's head, smooth the cover back across her body, and feed her hind legs through two leg straps — it's that easy! Rocky Sheep Suits (see Resources) also come in a two-piece, front Velcro-secured model that makes dressing horned sheep a snap.

Sheep covers are meant to be worn year-round, though some shepherds coat their sheep only during hay-feeding season. Elastic inserts allow the coat to expand a little as the sheep's fleece grows in, but most sheep need their covers exchanged for larger sizes at least three times a year.

You can certainly sew your own sheep covers if you want (see Resources), but breathable fabric is expensive and it's usually as cost-effective to buy ready-mades.

**THE ADVANTAGES:** Fleeces stay infinitely cleaner, the tips of colored fleece don't fade, and since all brands of sheep covers are white or tan, they reflect summer heat, keeping colored sheep in particular much cooler than their uncovered kin. And good covers are durable; most last at least four years.

**THE DISADVANTAGES:** You'll need an array of sizes to fit your sheep throughout the year, and you'll spend some time repairing rips and tears. Plan on having to catch your sheep and

change their covers when needed. It can be a job; sheep aren't keen on wearing clothing.

## More Ploys for Growing Better Fleeces

Even if you don't coat your sheep, you can grow better fleeces by following some basic guidelines, such as utilizing feeders that don't drop hay on your sheep's heads and by being careful not to accidentally fling hay or grain on sheep while feeding them. Another is to consider your bedding material. Chopped straw and woods chips are comfy for sheep to lie on but easily work their way into fleeces. Long-stem straw or the hay your sheep pull out of their feeders are much better selections.

A key step is carefully policing your pastures, paddocks, and barns to remove contaminants. Polypropylene baling twine and shredded fiber from poly tarps are major problems, as are certain natural plant materials, too. Have you ever tried to remove thistle debris or burdock burrs from a sheep's fleece? On or off the sheep, it nearly can't be done. Don't let these plants take over your pastures!

On your farm patrols, watch out for unusual contaminants that not only ruin fleece but injure sheep. One day our long-wool Wensleydale wether was hunching strangely as he walked along; when we checked, a small coil of old, rusty barbed wire was wrapped up in the fleece attached to one hind leg. Ouch!

The most important factor for high-quality fleeces, however, beyond keeping as much vegetable matter and junk out of your fleeces as possible, is to hire a competent shearer to shear your sheep.

---

### DID EWE KNOW?

#### Lanolin

Wool wax and wool grease were early names for lanolin (from Latin *lāna*, "wool," and oleum, "oil"). Lanolin was extracted by boiling wool in water and skimming the film of white grease left floating on top when the liquid cooled, then squeezing the grease through linen cloth to refine it.

It was used as-is for hand cream and blended with various herbs to formulate medicinal salves. To this day it's used in a myriad of products formulated for the protection, treatment, and beautification of human skin.

---

## Sheepshearing 101

Can you shear your own sheep? Certainly, especially if your flock is very small. The first four years we had sheep, we sheared our own. That's why I recommend finding a good shearer before you buy any sheep.

Although it's not as easy as it may sound: The worldwide dearth of sheepshearers becomes more acute every year. In 2008 sheepshearing was listed as one of Britain's most-needed occupations because shearers were so difficult to find. As older shearers retire, younger ones aren't there to take their place. It's hard work that requires a strong back and a good deal of skill.

When we were given our first two sheep, they hadn't been sheared in several years. Thinking "How hard can this be?," we ordered a good set of old-fashioned hand shears and a blade sharpener from Premier1. When they arrived at the farm, we set out to shear our sheep. Our first discovery: Sheep have exceedingly thin skin and it's easy to cut them. Second discovery: Removing the wool without ruining it looks infinitely easier than it is. By the time we finished, Dodger and Angel looked like they'd been ravaged by giant mutant moths.

The following spring, Baasha and her daughter Rebaa, Rebaa's daughter Shebaa, and our young ram, Abram, had joined our merry band. John thought we could shear them with our Oster A5 dog clippers. Because all of the sheep were low-grease breeds and we oiled the clippers roughly every two minutes, the A5s survived, but just barely. And the sheep still looked like mutant moth fodder.

The next year our flock had grown considerably, so we bought brand-new Oster electric sheep shears and lots of cutters and blades. It still took half an hour to shear a sheep and the roar of the bigger shears made the sheep go crazy. But we got their wool off and the sheep looked marginally better.

Summer set in early the following year and we had a lot of sheep by then. The first few sheep were shorn without too much sweat and tears (ours or theirs), and then came Gwydion. He was a sweet but very nervous wether who stressed out at the drop of a hat. We haltered him, tied his lead rope to the fence, and began shearing. He went bananas and resisted all efforts to calm him down. Ten minutes into the job, Gwydion fell over dead. We were stunned. John put up the shears and said never again.

We found Paul Ahrens, one of two shearers working the Ozarks, through an online directory of shearers. He was booked though early summer, but when he heard our plight, he graciously stopped by on his way to a relative's graduation party and sheared our sheep. He's come every spring since then. The man is worth his weight in gold.

The moral of this story is that you *can* shear, but if you do, learn to do it right. Buy the correct tools, be they hand or electric shears. Buy a fitting stand to secure your sheep and get them used to it before shearing day. Your best bet is to work with a shearer for a few days or to take a shearing course at one of the state universities that give them. But don't go at it blindly the way we did. It does not pay.

### Finding a Shearer

Start by checking listings in the sheepshearer directories at the American Sheep Industry Association and U.S. Sheep Breeders Online websites. Talk to your County Extension agent.

Ask the sheep specialist at your state university. Phone wool-sheep breeders in your area and ask who shears their sheep. Make a list and start calling shearers; do it three or four months before you want your sheep shorn. Their appointment books fill up early.

When you do find a shearer, he may not be willing to come to your farm to shear only a few sheep. Shearers charge by the head, usually $1.50 to $5 per sheep, so it isn't worth a shearer's travel time and expense, and the time spent setting up his equipment, to shear fewer than about 50 sheep.

In this case you can:

- Pay the shearer extra to cover his travel and set-up costs.
- Get together with other shepherds in your area to arrange for a local shearing day so the shearer can visit enough farms to make his trip worthwhile.
- Organize a shearing day at a central location — one specific farm or a fairgrounds,

perhaps — where small-scale shepherds can bring their sheep to be shorn. This is not an ideal situation, as it stresses your sheep and potentially exposes them to disease. If you go this route, provide your own sheet of plywood for the shearing floor and ask the shearer to change blades between each owner's sheep.

## The Bowen Technique
. . . . . . . . . . . . . . . . . .
Today's method of shearing sheep, the Bowen Technique, was developed by New Zealanders Godfrey and Ivan Bowen in the 1940s. They discovered they could shear quickly and accurately by moving a sheep through a series of positions that stretch her skin so the skin is less likely to be cut. The brothers' technique was eventually adopted by shearers all over the world.

Godfrey Bowen also promoted shearing as a spectator sport. He helped organize and then competed in the first Golden Shears national shearing tournament in 1961, where he finished second behind his brother. Winning numerous world champion shearer titles during his lifetime, Ivan celebrated his 90th birthday by shearing a sheep in 2.41 minutes flat!

(See illustrations of the Bowen Technique on pages 162 and 163 and photos on pages 62 and 63.)

## Get Ready to Shear

Set up a well-ventilated area with good lighting and a catch pen nearby. Put down a new (clean, flat, and snag-free) 4- by 8-foot (1.2 x 2.4 m) sheet of plywood to shear on. Buy or make a skirting table with a slatted surface; a small, close-mesh gate laid across two sawhorses works fairly well, as does a short length of welded-rod sheep panel fencing supported by two boards running along the long ends of the panel and sawhorses at both ends.

The working area should be big enough to hold two entire fleeces at a time. Set up the rest of your work area as follows:

- Stock water, soft drinks, and sandwiches in a cooler, and if it's cold outdoors, plan to serve coffee or other warm drinks, too.
- Buy a can of aerosol wound spray to spritz on cuts.
- Collect bags or boxes to hold fleeces. Paper feed sacks turned inside-out work well for small fleeces; clean, unwaxed cardboard boxes are good for fleeces of all sizes. Note: Do not use plastic bags.
- Arrange for helpers to catch sheep, sweep the shearing floor, carry fleeces to the sorting table, and to skirt and roll fleece.

Pen your sheep on the eve of shearing day. Keep them indoors, if possible, but don't bed them on straw or any other bedding that clings to fleeces. Wet sheep *cannot* be shorn — the fleece will mildew — so if they get wet overnight, you'll have to postpone shearing.

Don't feed your sheep any supper and remove their drinking water at daybreak. It's more comfortable for sheep to be tipped if their stomachs are empty.

Skirting table

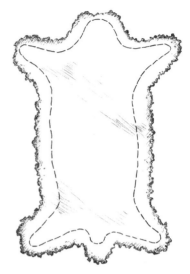

Where to skirt a fleece

## On Shearing Day

Be ready to go to work when the shearer arrives. Explain that you plan to sell your sheep's fleeces to handspinners, so it's important to avoid second cuts. (Second cuts are short, prickly bits of fleece created when a shearer goes over an already shorn area in an effort to neaten up a previously shorn area.)

Stay close to the shearer to advise him if any sheep need special treatment, such as pregnant ewes or old, arthritic sheep. If you have wethers, tell the shearer just before he shears them so he doesn't accidentally cut their penises.

Expect the shearer to nick your sheep; it happens. Sheep skin is very fragile and the best of shearers occasionally cut sheep. Be ready with your aerosol wound dressing to treat injuries before the sheep is released.

Sweep the shearing floor after each sheep.

## STEP-BY-STEP SHEEP SHEARING

While most shearers today use electric clippers (see photos on pages 62 and 63), some still prefer the old-fashioned human-powered variety. Most of them use the Bowen Technique (see page 160); some of the steps are shown here.

As each fleece comes off a sheep, a helper should carry it to the skirting table and energetically toss it down woolly side up and flesh side down, which dislodges dirt and the short pieces from any second cuts.

Whoever is manning the skirting table gives the fleece a good shaking, then pulls off manure-encrusted, stained, matted, or otherwise objectionable material from the edges of each fleece. Next she removes any remaining head, neck, leg, and belly wool. Finally, she determines which end is the neck and which is the butt, and then folds one-third of the right side of the fleece over the back and one-third of the left side over that. Starting at the butt, she rolls the folded fleece jelly-roll fashion up to the neck end, so that the prime shoulder wool is now on the top. She places the fleece in its own bag or cardboard box and moves on to the next one.

## You'll Roo the Day

*Rooing* is the process of removing the fleece from a sheep by hand-plucking the wool. All Soays and a high percentage of Shetland sheep shed the current year's fleece growth in late spring or early summer. By inserting spread fingers into the fleece of a sheep who sheds and working progressively along the sheep, it's possible to remove the fleece in one piece, exactly as though it was shorn.

During rooing, sheep can remain standing or they're rooed lying down with their legs tied or held together. Starting at one shoulder, the fleece is teased off until one side of the sheep is finished. The person doing the rooing then moves to the other side, or the sheep is turned over, and the other side completed.

It doesn't hurt a sheep to be rooed. Visit the Shetland Sheep Information website to view pictures and videos of rooing in progress.

## Washing (Scouring) Fleece

You may want to wash your fleece before storing or selling it. There are scores of acceptable ways to wash raw fleece. This one is based on instructions from the American Sheep Industry Association (ASIA).

You will need:

- A small- to medium-size stock tank or an old bathtub (you can do this in your regular bathtub but it's messy)
- A mild detergent that doesn't contain bleach. The ASIA recommends cheap powdered soap from Sam's Club, but soap contributes to unwanted felting, so it's not a best choice. Most fiber crafters recommend Dawn dish soap, but some swear by mild shampoo for humans or pets. There are products formulated specifically for washing fleece, but dish detergents seem to be most fiber workers' first choice.
- A thermometer to measure water temperature
- A perforated plastic crate of the type used to carry four 1-gallon milk jugs or to store files

- Several nylon mesh clothes washing bags, large enough to hold a washed fleece
- A drying table made of hardware cloth placed over sawhorses and supported by boards along the edges (a table is optional but nice to have)

Here is the process for washing fleeces:

1. Make sure the fleece you plan to wash has been properly skirted so that manure-laden parts have been removed. Pick off obvious vegetable matter and shake the fleece thoroughly to dislodge second cuts. Place the fleece in the plastic crate and pre-rinse it by repeatedly submerging it in hot (120°F [48.9°C]) water.

2. Fill your stock tank or bathtub with very hot water (160°F [71°C]). Add 1 cup of dish washing detergent and gently stir it in (a canoe paddle works extremely well, a piece of board will do), taking care not to raise suds.

3. Place the fleece in the water. Using your stirring implement, push it down to thoroughly saturate the wool. Let it remain in the tank for one hour; *don't agitate or otherwise disturb the fleece*. If you do, the fibers may start to felt and the fleece will be ruined. Remove the fleece by hand (you may want to wear insulated gloves) or with a pitchfork.

4. Rinse coarse or medium type fleeces in plain, warm (100–110°F [37.8–43.3°C]) water at least twice, changing water in the tank between rinses. If in doubt, repeat the wash cycle again before rinsing. Because it's much greasier, fine wool generally requires at least two more washes and two rinses.

5. When you're sure the fleece is clean and detergent-free, divide the fleece, placing it loosely in nylon mesh bags. Then place the bags in your washer and extract water from the fleece using the machine's spin cycle.

6. Spread the fleece very thinly over a drying table or on a clean floor. Turn the fleece two or three times until it's dry. If it feels at all tacky when it's dry, you'll have to wash it again.

## Storing Fleece

You can store raw fleeces in the same paper bags or cardboard boxes they were sorted into at shearing time, or in fabric bags made or recycled for this purpose. Bleached muslin bags are perfect for long-term storage and old pillowcases work well, too. *Do not* store raw or washed fleece in plastic bags or in poly feed bags; the storage containers have to breathe.

# Fiber Fests

A good way to learn (and have fun!) is to visit a sheep and fiber festival and take part in seminars and classes. Here are some of the biggest (websites listed in Resources).

**Black Sheep Gathering (Oregon)**
Festivities include sheep, fiber goat, wool, mohair, and fiber arts shows; scores of paid workshops and free demonstrations; a wool and mohair sale; and a huge vendors' marketplace.

**California Wool and Fiber Festival**
Part of the Mendocino County Fair and Apple Show. Festivities include the California National Wool Show, fiber arts judging, spinning contests, dog trials, demonstrations, and a large, active vendor's venue.

**Maryland Sheep and Wool Festival**
This granddaddy of American fiber fests celebrates its fortieth anniversary in 2013. There's a parade of sheep breeds; a plethora of demonstrations, workshops, and competitions; and more sheep, fleece, and vendors than you can shake a stick at. If you can only attend one festival, this is it!

**New York State Sheep and Wool Festival**
This is one of the largest and most prestigious sheep and fiber festivals in the world. Events include sheep, Angora goat, and fleece shows; Make It with Wool, knitting, and spinning contests; paid workshops and free demonstrations; fleece and sheep sales; and a huge vendor's area.

**Sheep Is Life Celebration (Arizona)**
Held on the vast Diné (Navajo) reservation, this festival features workshops on Navajo crafts such as rug and sash weaving, hand carding and spinning, native plant dying, and felting. Other events include a Navajo-Churro sheep show and sale, demonstrations of Navajo wool-working techniques, and auctions of weavings and rugs.

**Trailing of the Sheep Festival (Idaho)**
This event celebrates the annual tradition of moving thousands of sheep from summer high mountain pastures to winter pastures on the Snake River plain. It features fiber workshops, sheepdog trials, a large vendors' area, and a Lamb Dine Around of local restaurants. A connected folklife fair celebrates the Basque shepherds who historically tended the area's vast flocks.

**Wool Festival at Taos (New Mexico)**
In existence for 30 years, this celebration features — in addition to numerous demonstrations and workshops and a large vendor's area — a Fiber Critters Showcase, a fleece show, a handspun yarn competition, garment and home accessories judging, and some unique competitions such as speed crocheting and knitting and spinning the longest thread of yarn in one minute.

**Australian Sheep and Wool Show (Victoria)**
Highlights include a Festival of Lamb hosted by local eateries; the Women of Wool Gathering featuring wool-working workshops, demonstrations, and competitions; a fleece show; huge sheep shows; top-shelf Merino and Dorper sheep auctions; prestigious shearing and wool handling competitions; and sheepdog trials.

**Woolfest (Cumbria, England)**
This event is a great way to experience British wool firsthand. Woolfest is known for its large vendor's area and display of fiber-producing animals, including rare and local breeds, as well as plenty of experts, farmers, and breeders.

CHAPTER 12

# Got Milk?

*The mountain sheep are sweeter,*
*But the valley sheep are fatter.*
*We therefore deemed it meeter*
*To carry off the latter.*

— Thomas Love Peacock, *The Misfortune of Elphin* (1829)

Why milk a ewe? For the cheese and the yogurt and the ice cream! Sheep milk has considerably higher solids content than goat or cow milk has, so it makes a lot more cheese per gallon. Sheep milk has nearly twice the butterfat of cow milk; it's also richer in vitamins A, B, and E, calcium, phosphorus, potassium, and magnesium.

Rich sheep-milk cheeses and yogurt are an epicurean delight and you can make them yourself. Cheese making is an art you can learn from books. I recommend *Home Cheese Making: Recipes for 75 Delicious Cheeses* by Ricki Carroll and *The Home Creamery* by Kathy Farrell-Kingsley. They're the best!

Some of the world's great cheeses are crafted of sheep milk, so you can taste them before you commit. Visit your favorite gourmet cheese store and sample some of the cheeses from the following countries. Yum!

**BULGARIA:** Katschkawalj, Sirene

**FRANCE:** Roquefort, Perail, Abbaye de Belloc, Ossau-Iraty, Broccio

**GREECE:** Feta (can also be made of goat or cow milk), Kefalotiri, Myzithra, Kaseri Manouri

**HUNGARY:** Liptoi

**ITALY:** Pecorino Romano, Pecorino Sardo, Pecorino Toscano, Fiore Sardo, and Canestro Pugliese

**PORTUGAL:** Serra de Estrala

**ROMANIA:** Brinza, Teleme

**SPAIN:** Manchego, Zamorano, Roncal, Castellano, Idiazabal, Burgos, Villalon

**TURKEY:** Beyaz Peynir, Mihalic Peynir

## Choosing Your Dairy Ewe

A purebred or high percentage East Friesian, Lacaune, or British Milk Sheep will obviously give the most milk, but if your needs aren't great, nondairy ewes make great household dairy sheep. Too, don't discount a ewe you otherwise like because she isn't the "right" breed. If you like her, you *like* her. Factoring in the time you'll spend with her every day, that counts for a lot.

Most ewes milked in the United States are purebred or percentage East Friesians and East Friesian–Lacaune crosses. Canadian readers can also choose British Milk Sheep, which cannot be imported due to scrapie restrictions.

### EAST FRIESIAN

East Friesians are an old breed that originated in the Friesland region of northern Germany and the Netherlands, where they were historically kept in small numbers as household dairy providers. The East Friesian is said to be the heaviest milking dairy sheep in the world, with purebreds milking up to 1,100 pounds of milk during a 220- to 240-day lactation.

Most purebred East Friesians are white. Their heads and legs are wool-free, but they grow a heavy fleece of high-quality, 28 to 33 micron wool that is shorn prior to lambing. The breed is polled and has a short, naked tail. Rams weigh 225 to 275 pounds (102–125 kg); ewes are lighter at 160 to 180 pounds (72.5–82 kg). These docile sheep usually give birth to twins or triplets. The breed is not particularly hardy, so if you fancy East Friesians, you'll need good feed and a decent barn.

EAST FRIESIAN

### LACAUNE

The Lacaune is an ancient breed from southern France, where its milk is used for crafting Roquefort cheese. It produces less milk and births fewer lambs than do East Friesians but it's considerably hardier. Few purebred Lacaunes are found in Canada and the United States; however, dairy sheep of mixed East Friesian–Lacaune bloodlines are a favorite in American sheep dairies.

Lacaunes are white. They are polled and grow very little wool; their heads, legs, and

LACAUNE

most of their bellies are wool-free. Rams weigh 200 to 225 pounds (91–102 kg) and ewes, 150 to 170 pounds (68–77 kg).

## BRITISH MILK SHEEP

British Milk Sheep were developed in Britain during the 1970s. The exact genetic makeup of the breed is uncertain, but East Friesians, Bluefaced Leicesters, polled Dorsets, and Lleyns are among the parent breeds.

British Milk Sheep are nearly as milky as East Friesians and are known for lambing mostly triplets. They are polled, white with wool-free faces and legs, and they grow 9 to 14 pounds of 28 to 31 micron fleece per year. Rams weigh about 240 pounds (109 kg), ewes average 185 pounds (84 kg).

**BRITISH MILK SHEEP**

### DID EWE KNOW?

## Milk Mamas Rule!

Specialized dairy sheep produce 400 to 1,100 pounds of milk (that's 50–137½ gallons) per lactation while nondairy breeds give roughly 100 to 200 pounds (12½–25 gallons) per lactation.

# Milking Basics

Milking isn't for everyone. Make sure you're up to it before you commit to milking twice a day at the same time of day, day after day, week after week, with no respite until you quit or your ewe's lactation ends. If your ewe is a low producer or she's nursing lambs, once a day milking is doable but less productive. Finding a farm sitter willing to milk sheep is a formidable task. If you milk, you had better be a homebody or willing to schedule day trips and vacations around milking time.

Milking sheep require proper feed, and that means grain. Talk to your County Extension agent or an experienced sheep milker in your locale to formulate a nourishing diet, or feed alfalfa hay and commercially bagged sheep-specific grain, following the instructions on the bag.

Milking should take place in an area separate from your sheep's living quarters. A separate milking area is easier to keep tidy and prevents bedding, airborne dust, and flies from finding their way to the milk pail.

## Equipment Needed

You'll probably want a sturdy, goat-style milking stand. Whether you buy or build one, keep in mind that you'll be milking from behind instead of from the side as is done when milking a goat, so there should be enough room behind the ewe for you to sit on the milking platform or it should be short enough for you to get close while sitting on a separate milking stool. Also, unless your milking stand is very low to the ground, consider adding a ramp for your ewe to climb to the milking platform.

One way to set up your milking stand

Sheep aren't as agile as goats, so a ramp is easier for the ewe and it gives you a sturdy place to sit.

Alternatively, secure your ewe with a halter or collar and then squat or sit cross-legged on the ground behind her to milk her. It's worked for thousands of years!

In addition to a milking stand, you'll need the following:

- A small stainless steel pail or bowl to milk into (plastic receptacles can't be properly sanitized)
- A strip cup or dark-colored bowl
- Teat cleanser or unscented baby wipes
- Teat dip or an aerosol product such as Fight Bac

- A strainer that can hold 6½-inch (16.5 cm) milk filters
- A funnel
- Glass containers to store milk in the refrigerator

You also need plenty of soap and hot water to keep everything squeaky clean. Cleanliness is the key to great-tasting milk.

## Let's Milk a Ewe

In an ideal world you'd milk your ewe at 12-hour intervals. Because that isn't always possible, however, allow as much time between morning and evening milkings as you can. At

---

**LEARN FROM THE PAST**

In the earlier days of sheep keeping in Great Britain the milking of ewes and making of ewe cheese formed a part of the farming industry of the Cheviot Hills and border counties. The process began about the time the lambs were taken from their dams and continued for one to two months. The milking was done in the morning and was always performed by girls. Ewe cheese was highly esteemed as a stomachic, as well as a relish.

— William James Clarke, *Modern Sheep: Breeds and Management* (1907)

the appointed hour, scrub your hands, gather your pail and strip cup or dark-colored bowl, and head to the barn. Measure your ewe's grain ration and deposit it in the grain cup on your milking stand, then lead her to the milking area. She'll climb up on the stand and you'll fasten her neck in the stanchion.

1. Wipe off her belly and udder to dislodge loose dirt and bedding that you don't want to find in the milk pail. Place your milking stool behind the ewe.
2. Clean her teats and udder with a paper towel dipped in teat cleanser and wrung nearly dry, or use an unscented baby wipe.
3. Massage the udder for 30 seconds to facilitate milk letdown.
4. Aim several squirts of milk from each teat into the strip cup or bowl, one at a time, to test for abnormalities. Most bacteria is contained in the first few squirts, so these should be discarded in any case.
5. Place your milking container behind, not under, the ewe. Milk with both hands; or if you prefer, hold the container in one hand and milk with the other, alternating hands.
6. When you're finished, dip the end of each teat in teat dip or spritz them with Fight Bac, making certain the orifice (the opening in each teat) is coated.
7. Allow the ewe to hop down, take her back to the rest of your flock, and then grab your milking container and head to the kitchen. If you're milking more than one ewe, pour the milk into a communal stainless steel container, cover it (a hand towel works nicely), and proceed to the next ewe.

## MILKING TECHNIQUE

If you've never milked, start with just one teat. Don't include the udder itself in your grip — doing so hurts the ewe and ultimately damages her udder — simply wrap as many fingers as you can around the teat, leaving your pinky and any additional fingers suspended in air.

Gently nudge upward with the side of your hand to fill the teat reservoir, tighten your thumb and forefinger to prevent milk flowing back up into the udder, then successively tighten your middle ringer, ring finger, and pinky finger (if you're using it), forcing the milk out into your pail. *Never* pull the teat and *never* strip it between your thumb and forefinger; both practices are painful and damage the teats.

Proper way to hold the teat

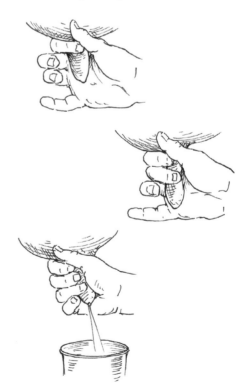

Milking seems difficult at first, and your back, fingers, and wrists will probably ache. Coordination, speed, and strength come with practice. In a week or so you'll be working alternate teats in a rhythmic, easy pace.

Milk until both teats are flaccid, then stop and gently massage the udder for 30 seconds to facilitate final milk letdown. Alternately, nudge the udder with the sides of your hands as you milk those last few squirts (to simulate lambs bunting their mama's udder).

Always use post-milking teat dip or Fight Bac, making sure the orifice of each teat is covered. Lambs suckling don't stretch the teat orifices very much, but milking by hand or machine does.

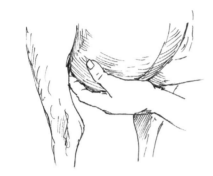

*Massage the udder for 30 seconds before starting to milk and again after the teats are flaccid to make sure all the milk is stripped out.*

*Always clean the udder before milking and dip the teats in an antibacterial solution when finished.*

## The Udder Truth

Ewes have one udder and two teats. The udder is her entire external mammary structure. You are not grasping her udders, as some folks mistakenly say, you are grasping only her teats.

Keep in mind that ewes' udders are not built like dairy goat udders. They're attached farther back between the ewes' legs, and they have smaller teats. Udder placement is why ewes are milked from behind instead of from one side as with goats and cows. You *can* milk from the side if you really want to, but it's a reach and hard on a milker's back.

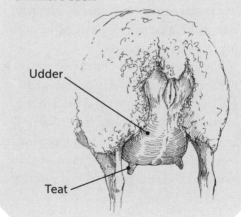

Udder

Teat

## Handling Milk

Milk should be processed as quickly as possible. Strain it through a milk filter into containers and for best results, submerge the containers in a sink of icy water for half an hour before refrigerating or freezing.

Sanitize your milking paraphernalia in hot, sudsy water, rinse thoroughly, and upend them in a drain rack to dry.

Never fail to examine those first squirts of milk directed into the strip cup; you're looking for stringy substances, blood, or anything unusual. If you find it, don't use the milk until you've tested for mastitis. Feed stores sell an inexpensive, do-it-yourself home mastitis testing kit called the *California Mastitis Test* (CMT) expressly for this purpose (see next page).

### Raw or Pasteurized?

Must you pasteurize the milk you drink? That depends. Some authorities think you'll be dead in weeks unless you do, while others claim raw milk is the healing ambrosia of the gods. The truth probably lies someplace in-between. Do your homework before you choose.

If you choose pasteurization, you can buy a home milk pasteurizer or pasteurize small batches on your kitchen range using one of the following methods.

To flash-pasteurize milk, place several inches of icy water in your sink. Fill the bottom of a double boiler with water and pour milk into the inner pan. Heat the milk to 161°F (71.7°C) (use a thermometer) and hold that temperature for 30 seconds, stirring constantly to prevent scorching. Then, set the inner pan in the sinkful of icy water and allow the milk to cool. This is the easy way. The downside: flash-pasteurized milk holds in the refrigerator only a relatively short time before it begins tasting "off."

You can also slow-pasteurize milk by gradually heating it to 145°F (62.8°C), then holding it at that temperature for 30 minutes, constantly stirring, before quick-cooling. Slow-pasteurized milk should stay fresh in the refrigerator for about one week, the same as fresh, raw milk.

## Sharing the Bounty

Some home dairy owners remove their ewes' lambs and raise them separately on lamb milk replacer. Most, however, allow lambs to nurse their mothers freely for their first two weeks of life and don't milk during that time. After that, the lambs are placed in separate quarters in the barn at night where they can see the ewes but not partake of their milk. If you choose this route (we do), feed your ewe in the evening but don't milk her. At the same time give her lambs their own little feed of hay and a smidge of grain. Milk your ewe in the morning, and then let her lambs rejoin her for the day. Lambs from dairy ewes are normally weaned around 8 weeks of age. After that, the milk is all yours.

### Easy as Cheese

I am not a kitchen wizard, but even I make delicious batches of these two easy cheeses (recipes on page 175). They're a springboard to more exacting recipes. Give them a try!

# Mastitis and the CMT

*Mastitis* is defined as inflammation of the udder. It's caused by a wide array of bacteria, including staph. Both intramammary infusions (introduced directly into the teat canal) and systemic antibiotics are generally used to treat this serious disease. If you think your ewe might have mastitis, have a milk sample cultured to find out which infective agents are involved; that way your vet will know which medications to prescribe.

Mastitis can also be caused by substandard milking hygiene and delayed milking, udder injuries, stress, and milk buildup after ewes wean their lambs. Untreated, it can progress to a more serious form called *gangrene mastitis* and, ultimately, death.

Mastitis isn't an everyday thing. If you take good care of your ewe, and milk under sanitary conditions, you'll probably never see mastitis in your barn. And if it does strike, you can clear up problems before they become acute by conscientiously testing your ewe's milk and treating subclinical mastitis.

Subclinical mastitis symptoms are subtle; they can be detected only in the strip cup or through testing. Acute mastitis symptoms include a hot, swollen, hard or lumpy udder; extreme pain; lameness; loss of appetite; fever; decreased milk production; clumps, strings, or blood in the milk; or watery, strange-looking milk. Gangrene mastitis presents as a bruised-looking, extremely swollen and painful udder that turns blue as infection takes hold.

Test for subclinical mastitis at weekly intervals. To do this you need a California Mastitis Test kit containing a plastic paddle with four compartments (the kit was designed for testing cows), a bottle of CMT solution that you'll reconstitute according to package instructions, and a test results card that shows precisely what you're looking for as you test.

You perform the test just before milking. Squirt 1 teaspoon (2 cc) of milk into each of two compartments of the paddle. Add an equal amount of reconstituted CMT solution to each compartment, and then rotate the paddle in a circular motion to mix the compartments' contents. Ten seconds after adding the solution, read the results (the visible reaction disintegrates after about 20 seconds, so don't delay).

Little to no thickening of the mixture occurs if the test is negative. Definite thickening indicates the possibility of mastitis; if the mixture gels, your ewe has mastitis.

If the CMT indicates mastitis or if you just aren't certain, take a sample of milk to your vet to be cultured. Don't wait to see if things get worse. This is not a disease you can ignore.

## Soft Citrus Cheese

. . . . . . . . . . . . . . . . .

½ gallon milk
⅛ to ¼ cup fresh lemon or lime juice
    Salt (optional)
    Herbs (optional)

**1.** Heat milk to between 175° and 180°F (77° and 82°C). Hold at that temperature and stir for 30 seconds, then slowly whisk ⅛ cup lemon or lime juice into the hot milk. Cover and let the milk set for 15 minutes until the curd clearly separates from the whey. If it hasn't separated at the end of 15 minutes, add small amounts of citrus juice until it does.

**2.** Pour the curds into a colander lined with butter muslin. Tie the corners in a knot and hang the bag to drain into a pan or the sink for several hours until no further liquid is seen coming out of the curds.

**3.** Remove the cheese from the bag. Add salt and herbs to taste. Store in a covered container in the refrigerator for up to 2 weeks. This cheese is delicious on crackers or bagels. You can also purée it in a food processor and add a dash of honey to make a tasty base for fruit dips.

## Queso Blanco (White Cheese, Panir)

. . . . . . . . . . . . . . . . .

½ gallon milk
⅛ to ¼ cup white vinegar
    Salt (optional)
    Herbs (optional)

**1.** Heat milk to between 175° and 180°F (77° and 82°C). Slowly whisk ⅛ cup vinegar into the hot milk. Hold that temperature, stirring constantly for 10 to 15 minutes, until curd clearly separates from the whey. If curd hasn't formed after 10 minutes, add additional vinegar a little at a time until it does.

**2.** Pour the curds into a colander lined with butter muslin and let cool for about 20 minutes. Tie the corners in a knot and hang the bag to drain for several hours until no further liquid is seen coming out of the curds.

**3.** Remove the cheese from the bag. Add salt and herbs to taste. Store in a covered container in the refrigerator for up to 2 weeks. Queso blanco is like tofu: it doesn't melt and it takes on the flavor of whatever it's marinated in or cooked with, making it a very versatile cooking ingredient.

**NOTE:** You can make a firmer, feta-like cheese from either of these recipes by pressing the cheese overnight. To press soft cheese, place it on a plate, cover it with another plate, and add a weight to the top (a couple of books or a big jar of peanut butter or mayonnaise works fine). Be sure the plate is large enough and curved a bit so the excess liquid doesn't spill over the sides. Or put the whole setup in the sink to drain.

### Freezing Excess Milk

Milk should be fresh from the ewe if you want to freeze it; milk that is refrigerated first tends to separate when thawed. To freeze, pour fresh raw or newly pasteurized milk into shatter-proof containers allowing 1 inch (2.5 cm) of headroom, chill in icy water, then pop them in the freezer for up to 6 months. We freeze raw milk in resealable quart bags and when frozen, double-bag in gallon-size bags. Thaw frozen milk slowly in the refrigerator to conserve its flavor and texture.

## Selling Milk

Don't sell milk, especially raw milk, before you know if it's legal. Selling milk from the farm is illegal in 21 states, and laws vary widely in the states that do allow it. Some states such as Florida, North Carolina, and North Dakota allow producers to sell raw milk "for pet food only." And in Kentucky and Rhode Island, customers can buy raw milk only with a doctor's prescription. Stiff penalties for trafficking in illegal milk exist. Know your legal rights before you sell it.

### This Milk Tastes *Nasty*!

Fresh, properly handled milk from healthy ewes is naturally creamy, sweet, and tasty. When it isn't, here are some things to consider.

- Milk from ewes who have mastitis tastes slightly strange to vile. When something isn't right, always test for mastitis before trying anything else.

- Improperly handled milk often develops off flavors, especially if it isn't quickly cooled after straining.

- Some types of feed may flavor milk if fed within 5 hours of milking time. These include alfalfa and sweet clover (green or baled into hay), green barley, soybeans in any form, rye, rape, and vegetables such as turnips, cabbage, and kale.

- Many plants, both wild and tame, impart unwelcome flavors. Plants known to taint milk include wild garlic and onions, bitterweed, mustard, marigolds, chamomile, fennel, ragweed, peppercress, wild lettuce, and Queen Anne's lace.

# Training Sheep

*Plainly the sheep and the wolf are not agreed upon a definition of liberty.*

— Thomas Jefferson

Those of you who have read *The Backyard Goat* will see that some but not all of the following material is reworded from that book. Sheep are every bit as smart as goats and can easily learn to pull a cart or wagon or perform the clever tricks I talk about in *The Backyard Goat*. If you want to train your sheep beyond what is discussed in this chapter, that book is a good starting place.

## Training Rams

It's not wise to train rams beyond basics like leading, standing tied, and picking up their feet; rams are unpredictable and shouldn't be encouraged to be overly friendly. Even if they seem to be easygoing, avoid squatting or sitting on the ground by them. You don't want to be bashed by a ram's powerful forehead.

## Taming Sheep

Before you can train a sheep, she has to be tame. Sheep are more fearful than goats and they are not wired to automatically tolerate human touch, so first you have to gain your pupil's trust. Take your trainee and a companion animal to a comfortable area where they can stay away from the main flock for a week or so.

- Start by teaching your sheep to eat from your hand. Plan to work with them several times a day.
- Go into their area and sit directly on the ground, a small stool, or an overturned feed or water tub, so that you're down at sheep's eye height. Bring a book or something to do and ignore them altogether the first day. Curiosity will bring them over to sniff you.
- On the second day, bring a feed pan and some grain in a small container. Place the pan in front of you, then ignore the sheep.

When they approach, still not looking at them, drizzle a handful of feed into the pan. Do this several times each visit.

- On the third or fourth day, depending on how trusting they seem, instead of drizzling the feed, hold it in your palm with the back of your hand against the bottom of the pan. Be patient. Eventually the sheep will venture a nibble. Hold your hand against the pan until the sheep readily eat from your hand. Then hold your hand at sheep nose level.

- Next, turn your sheep and her friend back in with the main flock. Walk among them later and offer the sheep you've hand-trained some grain. If she accepts it, your sheep is ready for further training.

- Now, at least once a day for several days, take your sheep to an enclosed area. Lead the flock with a bucket of feed, then turn them back out one at a time until she's the last one remaining. Talk to her quietly and feed her from your hand until she's comfortable being by herself. Now she's ready for training.

## Secondary Reinforcement

It's good to have a second way to reward your sheep before embarking on actual training. Most tame sheep adore having their chins, backs, and especially their chests scratched. Once your pupil confidently eats from your hand, quietly attempt to gently scratch her back or chest; it may take several tries, but once she discovers how nice this feels, she'll be putty in your hands.

## Training Basics

Any training system based on positive reinforcement works well with timid species such as sheep, but clicker training is the perfect way to go. Some people don't try clicker training because they think that once they do, their pupil will mob them for food. Quite the reverse is true. Once a sheep realizes that food appears only after she's heard a "well done!" click, she'll stop expecting handouts all the time. Once you begin clicker training, however, *never* indiscriminately feed treats from your hands. You can still treat your sheep to random goodies, but place them in a feed pan or on the ground and step away. Otherwise ask her to work for her treats. "Work" can be as simple as targeting on your hand, but expect her to earn handheld yummies.

Two other misconceptions are that once you've clicker trained an animal you'll always need a clicker to make her work, and that she'll work only for food. Clicks and rewards are used to teach *new* behaviors; once a pupil learns them, the clicker and food rewards are phased out and simple praise or a nice back scratch can take their place.

You don't have to buy a lot of gear to clicker train sheep. All you need is:

**A CLICKER.** Most pet stores sell them nowadays.

**A TARGET.** A target can be anything from an empty soda pop bottle to your hand. I like the handheld marine float type used by clicker-training guru Shawna Karrasch. Buy one, make your own (see page 180), or improvise.

**REWARDS.** In most cases this means food. Rewards should be something sheep love, broken into tidbits no bigger than a raisin. You're rewarding, not feeding, your pupil.

**A GOODIE BAG.** Stow rewards in a loose-fitting pocket or a separate receptacle, unless you're using food as a lure. Carpenter's aprons, the kind lumber yards hand out for free, make great goodie bags, as do fanny packs and horse- and dog-training reward bags that clip onto your pocket or belt.

**A QUIET, OUT-OF-THE-WAY SPOT** for training sessions, preferably enclosed, but in any case, away from other sheep.

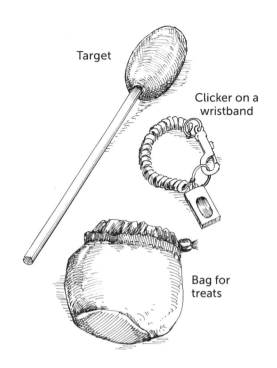

Target

Clicker on a wristband

Bag for treats

## Does Clicker Training Work for Sheep?

Though no one has written books or produced videos specifically about clicker training sheep, most of the material in dog- and horse-training books and DVDs applies to sheep as well. I recommend reading Peggy Tillman's *Clicking with Your Dog: Step-By-Step in Pictures* for starters.

If you doubt that sheep can be clicker-trained, visit YouTube and watch Portuguese dog trainer Fernando Silva train his lamb, Clarinha, over an agility course in the Clicker Sheep video. If Clarinha can do it, your sheep can, too!

## Choosing Rewards

Before you start your clicker-training project, discover what type of goodies your pupil likes best. Collect a selection of items, spread them out, then bring your sheep to where they are and let her choose.

If there is one item she loves above everything else, save that treat for jackpots. Jackpots are super-special rewards awarded for superoutstanding behavior. Jackpots can be a special treat or more than the usual amount of her typical training snacks. Don't hand them out indiscriminately — that's why they're jackpots.

Experiment with different treats, keeping in mind that most sheep have never been given "people food" and may be reluctant to try it. If that's the case, gently open her mouth, insert a goodie, and let her taste. She may spew the same item out several times before deciding it's really very good. Be patient. Here are a few things to try:

- Dry breakfast cereal (Cheerios are perfect)
- Animal crackers broken into smaller pieces
- Raisins or another dried fruit chopped into tiny bits
- Pelleted sheep feed, especially if it's not on her everyday menu
- Popped popcorn
- Packaged croutons (broken to size if needed)
- Peanut halves
- Sunflower seeds, shelled or not
- Pieces of tortilla chips or pretzels
- Chopped fresh fruit and vegetables such as apples, pineapple, pears, carrots, celery, or kohlrabi (this is messy; line your goodie pouch with a plastic bag)

And many sheep work well for nonfood rewards. For these, lavish praise or a scratch behind the ears or under the chin may be enough.

---

### Make Your Own Targets

I love the type of marine float targets that Shawna Karrasch sells on her On-Target Training website, but I make my own. Making them is the essence of simplicity:

1. Buy marine floats (you can get them at Walmart, where they come two to a package) and ½-inch (1.3 cm) dowel rod at a hardware or lumber store.

2. Cut the dowel rod to length, squirt a smear of superglue (or its equivalent) just inside both ends of the hole already drilled through one of the marine floats, add the handle, and there you are!

Make two: a short one for close-up work (mine has a 10-inch [25.4 cm] handle, made using a 15-inch [38 cm] piece of dowel rod) and a longer one for leading (the one I use has an 18-inch [46 cm] handle and is made of a 22-inch [56 cm] dowel rod); the long one can be shortened later if it's too unwieldy.

## Teaching Your Sheep about Targeting

By touching his nose to the target, your sheep learns to perform a task in order to earn a reward. Simple targeting leads to more complex maneuvers like leading quietly at your side, hopping up on a fitting stand when asked, or learning neat tricks. The most satisfying part of clicker training is seeing the lights go on the first time your pupil realizes what you want him to do. It's magic and you make it happen!

First, teach your sheep that a click means food. Click, then quickly hand him a tidbit of food. Sheep are smart; they quickly connect the click with the treat, so don't carry this to extremes. And don't try to hide the goodie bag. Show it to your sheep, jiggle it, let him sniff it. He'll probably nudge and nibble the bag — or you. As long as you're safe, try to ignore unwanted behaviors (standing on his hind legs with his front feet up on you would be an exception).

To teach your sheep to target, stand at his head with the target and clicker in your left hand (this is easier with a marine float target than with something bulky) and your right hand free to quickly dish out rewards. Hold the target close to his nose. He'll probably reach out to see what it is. The moment his nose brushes the target, click, reward with food, and pile on the praise. Then keep the target close to his nose; if he doesn't immediately touch it again, position it where he's apt to bump it when he moves his head. When he does, click, reward, and praise.

Keep it up until he pauses, considers, and then reaches out to touch the target on his own. Jackpot! Haul out the tastiest items in your goodie bag and celebrate. This may happen during your first session or it might take several, but once your pupil makes the connection, everything else is a breeze.

Once your sheep understands that touching the target elicits a reward, teach him to follow it with his nose. Start by moving it around within arm's length so he can touch it without shifting his feet. Then hold it up high, so he has to stretch and then near the ground, so he lowers his head. Finally, take a step back, so he has to step forward to reach it. Keep moving around in a small area until he understands that you want him to follow the target.

### The Ultimate Target

After your sheep has learned basic targeting behavior, train him to target on your hand. Once he does, it's easy to covertly lead him through intricate maneuvers and your target is always there when you need it!

*Sheep are just as smart as goats and can readily learn any number of tricks and behaviors.*

# The Ten Commandments of Clicking

**1.** Keep sessions short and fun. Two or three 5-minute sessions a day are more productive than a single 1-hour session.

**2.** Correct timing of the click is crucial. Picture this: You're trimming your sheep's hoof and she holds her leg up like a princess; you put the leg down, then click and reward. You think you're clicking for holding her foot up, she thinks you're clicking because she put it down. Click at the precise moment a desired behavior occurs, not seconds before or after.

**3.** Click once per correct behavior. If your pupil does something especially well, treat her to a jackpot reward but don't reel off a series of clicks.

**4.** Click for both voluntary and accidental movements toward your goal: this sets your pupil up for success. You can hold the target where her nose will bump it, coax or lure her into a position you want, or briefly place her there, but don't pull, push, or hold her in place; she needs to think she earned the reward of her own accord.

**5.** At first, don't hold out for perfect behavior. You can be pickier once she understands what she needs to do to earn a reward, but in the early stages click and reward for any effort that resembles the behavior you want.

**6.** Correct bad behavior by clicking good behavior. Mugging for food is a classic example: if your pupil nudges and nibbles the reward bag begging for food, ignore her. Look away or turn your back and count to five. Then give her a chance to *earn* her treat by targeting or performing some other simple task.

**7.** Don't reinforce undesirable behavior. If your sheep does something wrong or engages in spontaneous behavior during training (such as targeting when you didn't ask her to), stop what you're doing and count to five to let her know you're not going to reward that action.

**8.** Once you're sure your sheep understands a behavior, start clicking every second or third correct response, then stretch it out until you're clicking at irregular intervals. While it seems as though the sheep would find this discouraging, quite the opposite is true — she'll wonder which response will earn the reward and she'll try all the harder to get it.

**9.** Some sheep may need each process broken down into many small, rewardable steps before they understand what you want them to do. Take your pupil's prior history into consideration when planning training goals. A sheep who hasn't been handled very much won't respond in the same manner as a bottle baby.

**10.** If you get mad or frustrated, stop. Take a deep breath, then ask your pupil to perform a simple behavior you know she can do. Click and reward and then quit for the day. Don't link anger or tears with clicker training. Training is meant to be fun.

## Adding Cues

Cues are words or gestures used to request specific behaviors. For example, you could say "hoof" or "pick it up" to ask your sheep to pick up his foot.

Introduce a cue only after you've already trained your sheep to do the thing you're asking for and you're absolutely certain the behavior is going to happen. Then the cue becomes a stimulus for the behavior itself: the cue is followed by the behavior and the behavior is reinforced with a click and reward.

From this point on, click and reward only when your sheep performs the behavior after the cue, not when he does it on his own. That way he'll quickly learn that he only gets the click and treat if he responds to your cue.

When you're positive he understands the concept of targeting and when you're sure he'll do it 100 percent of the time, add a spoken cue as you offer the target (*touch* and *target* are logical choices). Once he understands, reward *only* when you ask him, via the cue word, to perform, and ignore spontaneous touches; then you're ready to shape behaviors.

## Teaching Your Sheep to Lead

When your sheep reliably follows a target and understands a verbal cue, teach her to walk on a loose lead at your side. Using your target and wearing your reward pouch on the right side of your body, stand on the left side of your haltered or collared sheep. Face forward with your shoulder even with the middle of your sheep's neck. The float part of the target should face away from the sheep.

When you're ready to begin, bring the float around in front of your pupil's nose about 18 inches (46 cm) ahead of her muzzle and give your verbal cue (*touch*). When she steps forward and touches the target, click, reward, and praise. Repeat this again and again until she understands what you want her to do. Next, hold the target in front of her and click as she takes a step. Then click for every two steps, then three.

Before long, she'll be following the target and you can increase the distance she travels before you click and reward. Introduce a new cue as she starts to walk forward (*walk*). When she understands, put the target away. Now you have a happy sheep walking calmly at your side.

## Picking Up Feet

Few things are more frustrating than wrestling with a sheep who doesn't want his hooves trimmed. To teach him to stand quietly for the process, start with your pupil standing on the ground or on a fitting stand. If he kicks or struggles when you touch his hooves, run your hand down each leg, clicking and rewarding until you reach a spot where he starts to object, then proceed more slowly, always clicking and rewarding positive responses until he's okay with you touching his hooves.

Next, lean into your sheep to shift his weight to his opposite leg; when he moves over, click and reward. Slick your hand down his leg, grasp his hoof and pick it up. Click while his foot is coming up, not as he shakes his leg or slams his hoof down again. If he starts to shake his leg or his hoof starts down before you have time to click, count to five without reinforcing the behavior and try again.

When he lets you pick up his foot and holds it up for even a second, click-reward (still holding his hoof off the ground) and set it back down again. Work toward three or more minutes of patient standing, slowly increasing the amount of time before he earns each reward.

You can teach your sheep to hop up on a fitting stand, pull a wagon, or perform tricks such as standing with his front feet on a pedestal. When you want to teach him something new, just bring out the target and begin.

## Join a 4-H Club

Sheep are a natural for the 4-H program, which since 1907 has been part of the United States Department of Agriculture (USDA). Originally created to provide farming education for rural youth, 4-H (which stands for Head, Heart, Hands, and Health) has grown into a national network of local clubs, featuring projects that range from computers to club lambs.

4-H projects are designed to teach young livestock keepers the best way to care for and enjoy their charges. These groups are run by volunteers, usually parents whose children have been involved with the program some time, under the direction of a County Extension agent. Leaders have had extensive experience with sheep and act as mentors for young sheep enthusiasts.

Programs are open to children age 9 to 19. Typical 4-H sheep projects teach children how to feed, care for, handle, and prepare their sheep or lambs for shows. Those interested in exhibiting their animals are encouraged to do so, usually at the county fair or county 4-H fair level, though winners go on to state and sometimes national competition. All 4-H members keep records on their projects and attend local meetings throughout the club year.

Sheep provide many years of fun and learning for children and adults who volunteer as leaders for sheep projects. All breeds of sheep can participate in breeding and showmanship classes. Although Hampshires and Suffolks dominate club lamb (meat breed) competition, other meat breeds can be shown as well. In some places, fleece projects are offered in which shorn fleeces are judged instead of sheep.

To find out how and when to join 4-H, talk to your County Extension agent. No matter what your interest is in sheep, if you're 9 to 19, join 4-H. It's fun!

QUADRUPEDS.

OVIS ARIES. COMMON SHEEP.

South Down Polled Sheep of the improved breeds. From the Stock of the late Duke of Bedford, Woburn.

Published as the Act directs Aug 1 1807, by Longman, Hurst, Rees & Orme, Paternoster Row.

# Glossary

## A

**abomasum.** The third compartment of the ruminant stomach; the compartment where digestion takes place.

**acquired immunity.** Resistance to disease acquired from a young animal's dam through her colostrum.

**afterbirth.** The placenta and any fetal membranes expelled from a ewe after kidding.

**anestrus.** The period of time when a ewe is not having estrous (heat) cycles.

**anthelmintic.** A substance used to control or destroy internal parasites; a dewormer.

**artificial insemination (AI).** A process by which semen is deposited within a female's uterus by artificial means.

## B

**banding.** Castration by the process of applying a fat rubber ring to a ram lamb's scrotum using a tool called an *elastrator*.

**black-faced breed.** A breed raised primarily for meat.

**bleating.** Sheep vocalization; also referred to as *calling*.

**bloat.** Excessive accumulation of gas in a ruminant's rumen.

**bolus.** A large, oval pill; also used to describe a chunk of cud.

**booster vaccination.** A second or multiple vaccinations given to increase an animal's resistance to a specific disease.

**bottle lamb.** A lamb raised on milk or milk replacer from a bottle instead of by her dam; also called a *bum, bummer, cade,* or *poddy* lamb.

**breech birth.** A birth in which the rump of the lamb is presented first.

**breed.** Sheep of a color, body shape, and other characteristics similar to those of their ancestors, capable of transmitting these characteristics to their own offspring.

**britch.** The lower thigh of a sheep. Britch wool is very coarse.

**bummer lamb.** A bottle lamb, usually refers to a bottle-fed orphan.

**bunting.** The act of a lamb poking her dam's udder to stimulate milk letdown.

**butting.** The act of a sheep bashing another sheep (or a human) with his horns or forehead.

## C

**carpet wool.** Strong or coarse and hairy wools used in the manufacturing of carpets.

**caseous lymphadenitis.** A contagious, incurable bacterial disease.

**cast.** When a sheep is unable to regain her footing, she is cast; unless rescued, she will suffocate. Also called *cast down*. In some places a cast sheep is called a *riggwelter*.

**castrate.** Removal of a male's testes.

**cattle panel.** A very sturdy large-gauge welded-wire fence panel; sold in various lengths and heights.

**CD/T.** Toxoid vaccine used to protect against enterotoxemia (caused by *Clostridium perfringens* types C and D) and tetanus.

**CL.** See *caseous lymphadenitis*.

**club lamb.** A lamb raised as a 4-H or FFA project.

**coccidiostat.** A chemical substance mixed with feed, bottle-fed milk, or drinking water to control coccidiosis.

**colostrum.** First milk a ewe gives after birth; high in antibodies, this milk protects newborn lambs against disease; sometimes incorrectly called or spelled *colostrums*.

**conformation.** An animal's body proportions and shape.

**cover.** To breed (a ram covers a ewe).

**creep feeding.** To provide supplementary feed to nursing lambs.

**crimp.** Uniform waviness in an individual lock of fiber.

**crook.** A staff with a hook at one end, used to direct and catch sheep.

**crossbreed.** An animal resulting from the mating of two entirely different breeds; also called a *crossbred*.

**crossbreeding.** Intentionally mating two or more breeds together as part of a breeding program.

**crutching.** Removing soiled wool from a ewe's hindquarters prior to lambing; also called *crotching*, *dagging*, or *tagging*.

**cud.** Undigested food regurgitated by a ruminant to be chewed and swallowed again.

**cull.** To eliminate an animal from a herd or breeding program; also the animal thus eliminated.

## D

**dag.** A manure-encrusted hank of wool, also called a *tag*.

**dam.** The female parent.

**dehorning.** The removal of existing horns.

**dental pad.** An extension of the gums on the front part of the upper jaw; it is a substitute for top front teeth.

**deworming.** The use of chemicals or herbs to rid an animal of internal parasites.

**disbud.** To destroy the emerging horn buds of a lamb by application of a red-hot disbudding iron.

**dock.** To shorten a lamb's tail.

**drench.** To give liquid medicine by mouth; also a liquid medicine given by mouth.

**drop spindle.** A handheld spike (known as the shaft) with a weight (a whorl) used for handspinning.

**dry ewe.** A ewe who is neither pregnant nor lactating.

**dystocia.** Difficulty in giving birth.

## E

**ear tag.** A plastic or metal identification tag attached to a sheep's ear using a tagging tool.

**elastrator.** A pliers-like tool used to apply heavy, O-shaped rubber bands called *elastrator bands* to a ram lamb's scrotum for castration.

**estrus.** The period when a ewe is receptive (she will mate with a ram; e.g., she is "in heat") and can become pregnant.

**estrus cycle.** The ewe's reproductive cycle.

**ewe.** A female sheep.

## F

**felting.** The ancient craft of using heat, moisture, and pressure to produce a non-woven sheet of matted material to produce wool felt.

**fiber.** Animal or plant matter used in making textile yarns and fabrics.

**fiber fineness.** Fiber diameter, usually expressed in microns.

**fleece.** The wool coat sheared from an individual sheep; also a synonym for wool.

**flehmen.** Curling of the upper lip in order to increase the ability to discern scent.

**flight zone.** The distance something scary can approach a sheep without the sheep running off.

**flock.** A group of sheep; sometimes called a *herd* or *mob*.

**flocking instinct.** The need of sheep to flock together as a group.

**fly strike.** When blowflies or bottle flies lay eggs in manure-encrusted wool or wounds and the eggs hatch into flesh-eating maggots.

**fold.** An overnight pen where sheep are kept safe from predators; also called a *yard*.

**forage.** Grass and the edible parts of browse plants that can be used to feed livestock.

**free-choice.** Available twenty-four hours a day, seven days a week; hay and mineral mixes are generally fed free-choice.

**freshen.** When a ewe lambs and begins to produce milk, especially applied to dairy sheep.

### G

**gestation.** The length of pregnancy.

**grading.** The sorting and classification of fibers according to cleanliness, staple length, strength, evenness, and fineness.

**grading up.** A means of creating sheep of purebred status by breeding a ewe of a different breed to a ram of the desired breed, then their daughters to another ram of the same breed, and so on until reaching the percentage required for purebred registration. Not all registries allow grading up.

**graft.** A procedure in which a lamb is transferred to and raised by a dam that is not her own mother.

**grease.** Unprocessed wool as it is shorn directly from the sheep and hasn't been washed or otherwise cleaned.

**guard hair.** Long, stiff, generally coarse fiber that projects from the softer undercoat of a double-coated sheep's fleece.

### H

**hair sheep.** A sheep with hair instead of wool on her body; some hair sheep grow a short fleece over winter but shed it the following summer.

**halter.** A rope or leather head harness used to restrain, lead, or tie a sheep.

**hank.** A 560-yard (512 m) unit of yarn wound on a reel.

**heart girth.** Circumference of the chest immediately behind the front legs.

**heat.** Estrus.

### I

**immunity.** Resistance to a specific disease.

**in lamb.** Pregnant.

**in milk.** Lactating.

**intramammary infusions.** Mastitis medicines inserted directly into a teat through its orifice.

### J

**jug.** An approximately 4- by 5-foot (1.2 x 1.5 m) pen where a ewe and her lambs are put for the first 24 to 72 hours after kidding.

### K

**ked.** A wingless fly that sucks sheep's blood.

**kemp.** Hard, brittle, opaque, medullated fiber found in the fleece of some sheep.

**ketones.** Substances found in the blood of late-term pregnant ewes suffering from pregnancy toxemia.

### L

**lactation.** The period when a ewe is giving milk.

**lamb.** (n.) A young sheep; also the meat of lambs. (v.) The act of a ewe giving birth.

**lanolin.** The greasy substance in wool.

**livestock guardian.** An animal who bonds with, stays with, and protects livestock: usually a dog, donkey, or llama.

### M

**market lamb.** A lamb raised to be sold as meat.

**mastitis.** Inflammation of the udder, which also affects the milk.

**medullated fiber.** Medullated fibers are true hair fibers mixed in with normal wool but which do not have the same spinning and dyeing properties as pure wool; usually found on the faces, head, and legs of wool sheep.

**milk letdown.** Release of milk by the mammary glands.

**milky.** Produces enough milk to enable multiple lambs to mature quickly.

**mothering pen.** See *jug*.

**mule.** A type of hardy, crossbred meat sheep created by crossing rams of specific breeds on ewes of different, specific breeds.

**mutton.** The meat of mature sheep.

## N

**nematode.** A type of internal parasite; a worm.

## O

**omasum.** The third part of the ruminant stomach; it's sandwiched between the reticulum and the abomasum.

**open ewe.** One who isn't pregnant.

**ovulation.** The release of an egg from the ovary.

**oxytocin.** A naturally occurring hormone important in milk letdown and muscle contraction during the birthing process.

## P

**paddock.** An enclosed area used for grazing.

**papers.** A registration certificate.

**parrot-mouth.** When the lower jaw is shorter than the upper jaw and the teeth hit in back of the dental pad; also called an *over-shot jaw*.

**pedigree.** A certificate documenting an animal's line of descent.

**percentage.** Partbred; a crossbred sheep who is at least 50% of a specific breed (for example, a percentage Dorper).

**placenta.** See *afterbirth*.

**polled.** A natural absence of horns.

**postpartum.** After giving birth.

**prepartum.** Before giving birth.

**purebred.** An animal of a recognized breed.

## R

**ram.** An uncastrated male sheep.

**raw wool.** Unwashed fleece exactly as it was shorn from the sheep.

**registered animal.** An animal who has a registration certificate and number issued by a breed association.

**reticulum.** The second chamber of a ruminant's stomach.

**roman-nosed.** The convex profile of breeds such as the Blueface Leicester.

**roo.** To remove the molting fleece of certain breeds by hand-plucking.

**rumen.** The first compartment of the stomach of a ruminant, in which microbes break down the cellulose in plants.

**ruminant.** An animal with a multicompartment stomach.

**rumination.** The process whereby a cud or bolus of rumen contents is regurgitated, re-chewed, and re-swallowed; "chewing the cud."

**rut.** The period during which a ram is interested in breeding ewes.

## S

**scouring.** Having diarrhea; also washing raw fiber to remove impurities such as dirt, sweat, and grease.

**scours.** Diarrhea.

**scrapie.** The sheep and goat version of "mad cow disease"; sometimes incorrectly spelled *scrapies*.

**scurs.** Horn buds that have broken the skin but not grown into horns.

**seasonal breeders.** Ewes who come into heat during only one part of the year; most sheep are seasonal breeders.

**second cuts.** Short pieces in a fleece caused by re-shearing over an already sheared spot.

**settle.** Become pregnant.

**shear.** To remove a sheep's wool.

**sheath.** The outer skin covering a male sheep's penis.

**shedding breeds.** Hair sheep breeds, such as Dorpers and Katahdins, who grow winter fleece but shed it without needing to be shorn.

**sire.** The male parent.

**skirting.** Removing stained or otherwise undesirable portions of a fleece.

**slaughter lamb.** A lamb produced specifically for the meat market.

**sow-mouth.** When the lower jaw is longer than the upper and teeth extend forward past the dental pad on the upper jaw; also known as *monkey-mouth* or *under-shot jaw*.

**standing heat.** The period during estrus (heat) when a ewe allows a ram to breed her.

**staple.** A lock of wool.

**staple length.** The length of a sheared lock of wool obtained by measuring it without stretching or disturbing the crimp.

## T

**tag.** A manure-encrusted hank of wool, also called a *dag*.

**tender wool.** Wool with weak places in the fibers caused by illness, exposure to weather, or poor nutrition.

**topknot.** Wool from the forehead of a sheep.

**total fleece weight.** The weight of an entire raw fleece.

## U

**UC.** Urinary calculi; mineral salt crystals ("stones") that form in the urinary tract and sometimes block the urethra of male sheep.

**udder.** The female mammary system.

## W

**wether.** A castrated male sheep.

**white-faced breed.** A breed kept primarily for wool production; colored individuals may have black faces but they're still considered members of a white-faced breed.

## Y

**yearling.** A sheep of either sex who is 1 to 2 years of age, or a sheep who has cut her first set of incisors.

# APPENDIX 1

## Common Mistakes

Here is a quick list of important things you should know about sheep; the hows and whys are explained elsewhere in this book. It was compiled with the help of good friends from my Hobby Farms Sheep Yahoo group (see Resources) — thanks, everyone!

1. Don't buy a single sheep unless it's a bottle baby bonded with humans. The smallest permanent flock should comprise at least three and preferably five sheep.
2. Don't isolate an individual sheep. Even when lambing, sick, or in quarantine, sheep should be within sight and sound of other large animals, preferably sheep.
3. Don't buy wool sheep unless you know you can hire a shearer or are willing to buy the necessary equipment and shear them yourself.
4. Don't underestimate how high and how far a determined sheep can jump.
5. Don't buy sheep at sale barns unless you want to bring home hoof rot, caseous lymphadenitis, and a whole lot more.
6. Build sheep-proof feed bins inside the feed room and keep the door tightly latched.
7. Don't abruptly change types or amounts of feed.
8. Moderate amounts of grain are okay but don't overdo it. Sheep were designed to eat forage (grass or hay).
9. Learn to recognize good hay, then buy considerably more than you think you'll need. Store it correctly, so it doesn't spoil.
10. Don't overfeed bottle lambs and don't scrimp on what you feed them.
11. Don't catch or lift sheep by their wool. It hurts them terribly and damages their flesh and skin.
12. Don't take on sheep unless you can protect them from predators, even your own dogs.
13. Find a sheep-savvy veterinarian before you buy sheep.
14. Find a local mentor or join sheep-oriented online groups. Assemble a reference library of sheep books.
15. Don't jump into full-fledged breeding right away.
16. If you breed, plan ahead. Don't breed unless you're willing to keep lambs until they find good homes or to market excess lambs for meat.
17. Don't confine horned sheep behind large-mesh fencing; they'll get their heads or horns stuck and you'll have to cut them out.
18. Don't stress sheep any more than you have to.
19. Don't underestimate a sheep's sex drive. They will breed through gates and fences, so don't keep rams and ewes in adjacent paddocks.
20. Finally, don't keep a ram unless you need one. If you keep a ram, don't ever, *ever* take him for granted.

### Trimming Hooves

Trimming is the most important thing you can do to keep hooves in healthy condition. Soil moisture and type, time of year, and breed all influence how fast hooves grow, but as a rule of thumb, plan to trim hooves at least three or four times a year. Timing them to coincide with other labor-intensive procedures such as deworming or vaccinating makes sense. However, avoid trimming hooves during stressful periods such as shearing, late pregnancy, or weaning. Hooves trim more easily when they're moist, whether from dew, rain, or snowmelt.

You'll need the proper tools to do the job easily and correctly. Most shepherds use hoof shears; they're inexpensive, easy to handle, and tailor-made for the job. Though not absolutely necessary, a hoof plane or hoof rasp (buy the kind designed for miniature horses) is nice to have to level the foot when you're finished.

To trim your sheep's feet, first halter and tie him to something solid or jump him up on a milking or fitting stand, then you can squat beside him, perch on an overturned bucket, or stand and lean over him to trim the feet on

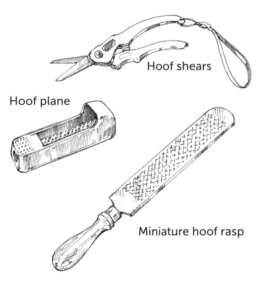

Hoof shears

Hoof plane

Miniature hoof rasp

### Goat Owners Take Note!

If you're used to trimming goat hooves, proceed with caution when you trim your first sheep. It's much easier to *quick* (trim the hoof so deeply that it bleeds) the soles of sheep than it is goats. And sheep bleed more profusely than goats, so keep Blood Stop powder or flour close at hand to staunch bleeding if you trim too far.

### The Perfect Hoof

To know how to trim sheep's hooves, examine the hooves of a 2- to 3-week-old lamb from the bottoms and from the sides. Trim at those angles and proportions; that's an ideal hoof.

the side opposite from the one where you're standing. Or you can set him up on his butt and bend forward to trim his feet.

Clean off the hoof with a brush or your hand, then begin trimming at the heel and work forward. Trim ragged outside edges even with the sole (the soft, central portion of each toe). Next, trim the inside walls level to match. If you have a hoof plane or rasp, carefully plane the bottom of each claw. When you're finished, the hoof should be flat on the bottom and parallel or nearly parallel to the coronary band.

If you cut too deep and a hoof bleeds, staunch the bleeding with Blood Stop powder or flour, then treat it with a wound treatment such as Blue Kote. Be sure to monitor the sheep for the next week or so to make certain it doesn't get infected. If the cut is a bad one, give the sheep a tetanus booster or a shot of tetanus antitoxin. Then, fit the sheep with a rubber boot made for the purpose (sheep and goat suppliers like Premier1 and Hoegger's carry them) or cover the bottom of the hoof with a thick layer of gauze pads and secure them with self-sticking, disposable tape such as VetWrap, running part of the bandage under the hoof and making sure you don't wrap it too tightly. Keep the sheep on soft pasture or indoors and replace the covering every day until the sheep stops limping; this keeps irritants out of the wound.

**NOTE:** When trimming a sheep with foot disease, trim the bad hoof or hooves last. Otherwise, you'll spread disease to his healthy hooves. When you're finished, disinfect your tools to avoid infecting the rest of your sheep.

# APPENDIX 3

## Promoting Your Sheep

If you want to earn money with your backyard sheep, you have to promote them. Here are a few suggestions adapted from *Storey's Guide to Raising Miniature Livestock*. Let some of them work for you!

- If you raise lambs to sell for meat, serve homegrown lamb whenever and wherever you can, especially at public functions and to people who may not have tasted it before. Distribute recipe sheets with your contact information printed on them.
- If you sell expensive, registered sheep, shoot good video footage of your animals. Edit it well, add music if you like, and burn it to CD or DVD. Distribute copies for free through your website or at your booth at shows and fairs.
- Show your fleeces at fiber shows. Create a booth to take to festivals and fairs. Demonstrate carding, spinning, or felting, using your own beautiful fleece.
- Put up a large, legible road sign by your driveway. If you live off the beaten path, provide directional signs with your farm name on them.
- Take your sheep out where people see them. Show them. Take a friendly sheep or bottle lamb to hospitals and nursing homes to visit shut-ins. Give talks and demonstrations. Prepare an educational program and take it to area schools. Volunteer to speak at seminars. Hold an open house. Sponsor or lead a 4-H project.

- Read stories to Head Start or kindergarten classes from books featuring sheep. Take along a living example. Tip off your local newspaper in advance.
- Participate on e-mail lists; be the first to field questions. People buy from people they know and respect, and you can advertise on many lists for free.
- Upload videos to YouTube. Show off your ram, teach viewers how to skirt a fleece, demonstrate how to teach a show sheep to lead. End each video with a short talk about your farm, including contact information.
- Blog. Add new content every week.
- Add a business-related signature to your outgoing e-mail; incorporate your farm name, location, a tag line describing your services, and your farm's Web address.
- Write an educational column for an area newspaper. Write it for free with one caveat: your contributor's byline must include your farm's contact information.
- Write reports and e-books about fiber arts or about training, breeding, or showing the breeds of sheep you raise; include a page about your business, of course. Publish them in PDF file and offer them as downloads through your website. Visit my Dreamgoat Annie website for samples; they're free!

# APPENDIX 4

## Emergency Euthanasia

Sometimes the unthinkable happens and a sheep or lamb becomes so seriously ill or injured that she must be euthanized. If a veterinarian can come right away and handle the task, so much the better; if he can't, you might have to do the deed yourself. The most acceptable way is to shoot the sheep. It sounds distressing and it is, at least for the humans involved — but we've done it, and the sheep's suffering ended quickly instead of her lingering for hours until the vet could stop by.

If you don't plan to eat the sheep and your vet has entrusted you with prescription drugs such as Rompun or Xylazine, give a hefty dose to sedate her (discuss this with your vet in advance, so that in an emergency you can administer the dose he recommends). If she relaxes so much that she lies down, so much the better.

Use a .22-caliber long rifle, a 9 mm or .38-caliber handgun, or if nothing else is available, a shotgun 410 gauge or better with a rifled slug. The muzzle should be held at least 4 to 12 inches (10–30 cm) away from the sheep's skull when fired. Using hollow-point or soft-nose bullets increases brain tissue destruction and reduces the chance of ricochet.

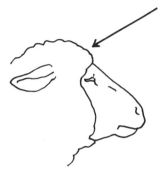

Aim from the midline of the back of a horned sheep's skull, just behind the bony ridge where the horns protrude, toward the back of the animal's chin.

Polled sheep and horned lambs less than 4 months of age should be shot from the same vantage point or from the front by aiming at a point ½ inch (about 1.5 cm) above the intersection of a diagonal line drawn from the base of the ear to the inside corner of the opposite eye. The firearm should be held perpendicular to the front of the skull.

When performed skillfully, euthanasia by gunshot induces immediate unconsciousness, is inexpensive, and does not require close contact with the animal. This method should be attempted only by individuals trained in the use of firearms and who understand the potential for ricochet. All humans and other animals should stay out of the line of fire.

# Resources

## Sheep Organizations

**American Association of Small Ruminant
  Practitioners**
Montgomery, Alabama
334-517-1233
*www.aasrp.org*

**American Sheep Industry Association**
Englewood, Colorado
303-771-3500
*www.sheepusa.org*

**Black and Coloured Sheep Breeders'
  Association of New Zealand**
Duvauchelle, New Zealand
+64-03-304-5090
*www.colouredsheep.org.nz*

**British Coloured Sheep Breeders Association**
Cross Ash, United Kingdom
*post@bcsba.org.uk*
*www.bcsba.org.uk*

**British Sheep Dairying Association**
Horsted Keynes, United Kingdom
*office@sheepdairying.co.uk*
*www.sheepdairying.co.uk*

**British Wool Marketing Board**
Bradford, United Kingdom
*mail@britishwool.org.uk*
*www.britishwool.org.uk*

**Canadian Sheep Breeders' Association**
Bluffton, Alberta
866-956-1116
*www.sheepbreeders.ca*

**Canadian Sheep Federation**
Guelph, Ontario
888-684-7739
*www.cansheep.ca*

**Dairy Sheep Association of North America**
Strum, Wisconsin
715-695-3617
*www.dsana.org*

**National Sheep Association**
Malvern, United Kingdom
*enquiries@nationalsheep.org.uk*
*www.nationalsheep.org.uk*

**Natural Colored Wool Growers Association**
Darlington, Pennsylvania
*ncwga@accuregister.com*
*www.ncwga.org*

**Navajo Lifeway, Inc.**
Window Rock, Arizona
505-368-3975
*www.navajolifeway.org*

**New Zealand Sheepbreeders' Association**
Christchurch, New Zealand
+64-03-358-9412
*www.nzsheep.co.nz*

## Rare Breed Organizations

**American Livestock Breeds Conservancy**
Pittsboro, North Carolina
919-542-5704
*www.albc-usa.org*

**Rare Breeds Canada**
Notre-Dame-de-l'Île-Perrot, Quebec
*rbc@rarebreedscanada.ca*
*www.rarebreedscanada.ca*

**Rare Breeds Conservation Society
  of New Zealand**
Christchurch, New Zealand
*enquiry@rarebreeds.co.nz*
*www.rarebreeds.co.nz*

**Rare Breeds Survival Trust**
Stoneleigh Park, United Kingdom
*enquiries@rbst.org.uk*
*www.rbst.org.uk*

**Rare Breeds Trust of Australia**
Abbotsford, Australia
+61-04-0832-4346
*www.rbta.org*

# Recommended Websites

**Cornell Sheep Program**
Cornell University
*www.sheep.cornell.edu*
    Links to a vast amount of information about sheep

**Department of Primary Industries**
New South Wales, Australia
*www.dpi.nsw.gov.au*
    Click on *Agriculture NSW*, then *Livestock*, and then *Sheep* to access a vast number of downloadable resources.

**Fias Co Farm**
*www.fiascofarm.com*
    This is my favorite dairy goat site; it's a best-bet source for information on kidding, home dairying and cheese making, and health care issues, nearly all of which can be applied to sheep and shepherding.

**Maryland Small Ruminant Page**
*www.sheepandgoat.com*
    The most extensive sheep and goat links on the Internet, this is the first place I check when I need information.

**National Scrapie Educational Initiative**
National Institute for Animal Agriculture
*www.eradicatescrapie.org*
    USDA scrapie laws are confusing; visit for a fairly clear treatment of the subject

**Navajo Sheep Project**
*www.navajosheepproject.com*
    Click on *Sheep Sheets* to download close to 100 well-written, educational, sheep sheets under nine headings

**Ontario Ministry of Agriculture, Food and Rural Affairs**
*www.omafra.gov.on.ca*
    Click on *Agriculture*, then *Livestock*, and then choose from a huge selection of sheep- and farm-related articles, some downloadable as PDF files

**Pipestone Veterinary Clinic**
*www.pipevet.com*
    Scores of educational pictures and articles, a sheep newsletter, and an "ask a vet" service, plus a full range of sheep supplies for sale

**Sheep 101**
*www.sheep101.info*
    Susan Schoenian's excellent introduction to all things sheep

**Sheep 201**
*www.sheep101.info/201*
    A natural progression from Sheep 101

**Sheep at Purdue**
Purdue University
*www.ansc.purdue.edu/SH*
    Links to hundreds of university-generated sheep bulletins

**Sheep Home Study Course**
Penn State Extension
*www.extension.psu.edu/courses/sheep*
    Take this comprehensive shepherding course (for noncredit) for free

**Shepherd's Notebook**
*www.mdsheepgoat.blogspot.com*
    Written by Susan Schoenian, Sheep and Goat Specialist for University of Maryland's Cooperative Extension

**U.S. Sheep Breeders Online Directory**
*www.nebraskasheep.com/directory*
    The most extensive sheep-related directory on the Internet with more than 3,500 entries

# Sheep-Oriented Yahoo Groups

**Hobby Farm Sheep**
*http://pets.dir.groups.yahoo.com/group/HFSheep*
    This is my list, please join!

**The Sheep Group**
*http://pets.dir.groups.yahoo.com/group/TheSheepGroup*
    A beginner-friendly sheep group

**Totally Natural Goats**
*http://groups.yahoo.com/group/TotallyNaturalGoats*
    This fantastic goat group encourages sheep owners to participate

## Periodicals

*The Banner Sheep Magazine*
309-785-5058
*www.bannersheepmagazine.com*

*Black Sheep Newsletter*
503-621-3063
*www.blacksheepnewsletter.net*

*Sheep!*
800-551-5691
*www.sheepmagazine.com*

*Sheep Canada*
888-241-5124
*www.shepherdsjournal.com*

## Supplies

**American Veterinary Identification Devices, Inc. (AVID)**
Norco, California
800-336-2843
*www.avidid.com*
A leading manufacturer of pet and livestock microchips

**Gentle Spirit Behavior & Training**
Olympia, Washington
360-438-1255
*www.gentlespiritllamas.com*
Cathy Spalding's Gentle Spirit halters are perfect for sheep

**Good Shepherd Lamb Coats**
Huntingtown, Maryland
301-643-0027
*www.goodshepherdlambcoats.com*
The crème de la crème of lamb coats, made out of felted, recycled, all-wool blankets

**Hoegger Supply Company**
Fayetteville, Georgia
800-221-4628
*www.hoeggerfarmyard.com*
Everything a sheep owner might need!

**HomeAgain**
888-466-3242
*http://public.homeagain.com*
Microchips; mainly for pets but they work for sheep and other livestock

**Jeffers, Inc.**
Dothan, Alabama
800-533-3377
*www.jefferslivestock.com*
A general line of supplies for all types of livestock, including wormers and vaccines

**Karen Pryor Clicker Training**
Waltham, Massachusetts
800-472-5425
*www.clickertraining.com*
Clickers, targets, training gear, books; click on *Library* while you're there

**On-Target Training by Shawna Karrasch**
Encinitas, California
800-638-2090
*www.on-target-training.com*
Clickers, targets, books; information

**Port-a-Hut**
800-882-4884
*www.port-a-hut.com*
Quonset-style portable housing for sheep

**Premier1 Supplies**
Washington, Iowa
800-282-6631
*www.premier1supplies.com*
Sheep and goat supplies, fencing, and shearing supplies

**Sera, Inc.**
Central Biomedia, Inc.
Shawnee Mission, Kansas
913-541-1307
*www.seramune.com*
Manufactures Goat Serum Concentrate, an acceptable IgG replacement product for sheep

**Sheepman Supply Co.**
Frederick, Maryland
800-331-9122
*www.sheepman.com*
  An assortment of sheep supplies including lamb
  coats and year-round sheep covers with elasticized
  bottoms

**Valley Vet Supply**
Marysville, Kansas
800-419-9524
*www.valleyvet.com*
  A general line of supplies for all types of livestock,
  including de-wormers and vaccines

**Woolover Limited**
Christchurch, New Zealand
*woolover@xtra.co.nz*
*www.woolover.co.nz*
  Woolen lamb covers made in New Zealand but sold
  by North American agents

## Sheep Shearers

**Shearer Directory**
American Sheep Industry Association
*www.sheepusa.org/Shearer_Directory*

**The Shearing Network**
*www.shearingnetwork.com*

**Sheep Shearers**
U.S. Sheep Breeders Online Directory
*www.nebraskasheep.com/directory/
  Sheep_Shearers*

## Predator Protection

**Livestock Guardian Dogs**
*www.lgd.org*
  The definitive livestock guardian dogs website

**"Livestock Guardian Dogs: Protecting Sheep
  from Predators"**
Animal Welfare Information Center, National
Agricultural Library
*www.nal.usda.gov/awic/companimals/guarddogs/
  guarddogs.htm*
  An Agriculture Information Bulletin, #588, by
  Jeffrey S. Green and Roger A. Woodruff, 1999. This
  USDA publication is a must-read for anyone con-
  sidering a livestock guardian dog.

**"Predator Control for Sustainable and Organic
  Livestock Production"**
National Center for Appropriate Technology
*https://attra.ncat.org/attra-pub/viewhtml.
  php?id=189*
  Written by NCAT staff, 2002, this is the best all-
  around guide to predator control; read it online or
  download it as a PDF file.

**South East Llama Rescue**
*www.southeastllamarescue.org*

## Dairying and Cheesemaking

**"Dwarf Goat Milk Stand Plans"**
*Dairy Goat Journal*
*www.dairygoatjournal.com/issues/85/85-3/
  Melissa_Thomas.html*
  Written by Melissa Thomas and featured in the
  May/June 2007 issue of *Dairy Goat Journal*. The
  step-up feature gives the milker a place to sit; mate-
  rials can be adjusted for larger sheep.

**Fias Co Farm**
Okemos, Michigan
*www.fiascofarm.com*
  Learn to make dairy products, plus find plans to
  make a milking stand.

**"Goat Milk Stand — Metal"**
Extension Publications, University of Tennessee
*www.bioengr.ag.utk.edu/extension/extpubs/
  Plans/6399.pdf*
  Building plans for a metal goat milk stand from the
  University of Tennesee's Cooperative Extension

**Hamby Dairy Supply**
800-306-8937
*www.hambydairysupply.com*
  Dairying supplies

**Hoegger Supply Company**
Fayetteville, Georgia
800-221-4628
*www.hoeggerfarmyard.com*
  Dairying supplies

**Leeners**
Northfield, Ohio
800-543-3697
*www.leeners.com*
  Cheese making kits plus a vast amount of do-it-yourself information

**"Making Homemade Cheese"**
Publications & Videos, College of Agricultural, Consumer and Environmental Sciences
*http://aces.nmsu.edu/pubs/_e/E-216.pdf*
  A New Mexico University Cooperative Extension Service Guide, E-216, written by Nancy C. Flores, 2008. All the basics from New Mexico State University extension

**New England Cheesemaking Supply Company**
South Deerfield, Massachusetts
413-397-2012
*www.cheesemaking.com*

# Fiber

**Fiber: Wool, Mohair, and Cashmere**
Maryland Small Ruminant Pages
*www.sheepandgoat.com/fiber.html*
  Extensive links to information about custom processors, do-it-yourself processing, fact sheets, fiber evaluation and testing, shearing, marketing, and more

**Handspinner**
*www.handspinner.co.uk*
  Don't miss the how-to videos and book reviews at Handspinner U.K.

**How to Make Sheep Coats for Your Own Sheep**
Gleason's Fine Woolies Ranch
*www.gfwsheep.com/sheepcoats/sheep.coats.html*
  Detailed information on making sheep covers

**The Joy of Handspinning**
*www.joyofhandspinning.com*
  A stellar source of articles and instruction videos

**MacAusland's Woollen Mills**
Bloomfield, Prince Edward Island
877-859-3005
*www.macauslandswoollenmills.com*
  Send your raw or washed wool to MacAusland's to be woven into wonderful, lightweight woolen blankets and lap robes. We sleep under MacAusland blankets made from our sheep's wool and we love them!

**Making Sheep Coats**
Desert Weyr
*www.desertweyr.com/sheep/sheepcoats.php*
  Sew your own!

**Matilda Sheepcovers**
Sydney, Australia
*sales@sheepcovers.com.au*
*www.sheepcovers.com.au*
  Sheep covers from Australia, available from North American agents

**Montana Wool Laboratory**
Montana State University
Bozeman, Montana
406-994-2100
*http://animalrange.montana.edu/facilities/ woollab.htm*
  Wool testing, even for nonresidents, at very low prices

**Rocky Sheep Company**
970-622-9965
*www.rockysheep.com*
  American-made sheep covers, including a horned sheep version

**Spinning Daily**
Interweave Press, LLC
*www.spinningdaily.com*
  Blogs, forums, videos, free PDF downloads of articles from *Spin-Off* magazine, and a guide to spinning guilds in every state

### Wool Festival

*www.woolfestival.com*

An online source for information about fiber events and festivals, free patterns, articles, and videos on spinning, knitting, weaving, felting, and more!

### Yocom-McColl Wool Testing Laboratories, Inc.

Denver, Colorado

303-294-0582

*www.ymccoll.com*

Tests raw wool for yield and fiber diameter

## YahooGroups and Periodicals

### SheepThrills

*http://dir.groups.yahoo.com/group/SheepThrills*

### Spin-Off

Interweave Press, LLC

800-767-9638

*www.spinningdaily.com/blogs/spinoff*

### Spindlers

*http://dir.groups.yahoo.com/group/spindlers*

### Spinning

*http://groups.yahoo.com/group/Spinning*

### The Spinning Wheel Sleuth

978-475-8790

*www.spwhsl.com*

### Wild Fibers

207-594-9455

*www.wildfibersmagazine.com*

# Miscellaneous

### Biodiversity Heritage Library

*www.biodiversitylibrary.org*

Enter *sheep* in the search box to download any of hundreds of vintage livestock books in free PDF format.

### The Freecycle Network

*www.freecycle.org*

The best place to get used mineral lick feeders and watering containers locally and for free!

### Google Books

*http://books.google.com*

Enter *sheep* in the search box, then click on *Free Google eBooks* to download the complete text of scores of vintage sheep books.

### "How Much Does Your Animal Weigh?"

*Backyards & Beyond*, Arizona Cooperative Extension

*http://ag.arizona.edu/backyards/articles/ winter07/p11-12.pdf*

An article on how to tape weigh sheep, goats, cattle, horses, and pigs by Susan Pater and featured in Arizona's Cooperative Extension's magazine, *Backyards & Beyond* (Winter 2007)

# Recommended Reading

+ Indicates vintage books that are in public domain and may be downloaded as free PDF files at Google Books and the Biodiversity Heritage Library.

+ Allen, R. L. *Domestic Animals: History and Description of the Horse, Mule, Cattle, Sheep, Swine, Poultry, and Farm Dogs.* New York: Orange Judd & Co, 1847.

+ Anonymous. *Twenty-Seventh Annual Report of the Bureau of Animal Industry 1910.* Washington, D.C.: Government Printing Office, 1912.

+ Bailey, L. H., ed. *Cyclopedia of Farm Animals.* New York: Macmillan Company, 1922.

Bradbury, Margaret. *The Shepherd's Guidebook.* Rodale Press, 1977.

+ Canfield, Henry J. *The Breeds, Management, Structure and Diseases of the Sheep.* Aaron Hinchman: Salem, OH, 1848.

Carroll, Ricki. *Home Cheese Making: Recipes for 75 Homemade Cheeses,* 3rd ed. Storey Publishing, 2002.

———. *Making Cheese, Butter & Yogurt.* A Storey Country Wisdom Bulletin, A-283. Storey Publishing, 2003.

+ Clarke, William James. *Modern Sheep: Breeds and Management.* Chicago: American Sheep Breeder Co., 1907.

+ Coleman, John, ed. *The Cattle, Sheep and Pigs of Great Britain.* London: Horace Cox, 1887.

Dalton, Clive, and Marjorie Orr. *The Sheep Farming Guide: For Small and Not-so-small Flocks.* Hazard Press, 2004.

Damerow, Gail. *Fences for Pasture & Garden.* Storey Publishing, 1992.

Dawydiak, Orysia, and David Sims. *Livestock Protection Dogs: Selection, Care and Training,* 2nd ed. Alpine Publications; 2004.

Dohner, Janet Vorwald. *Livestock Guardians: Using Dogs, Donkeys, &Llamas to Protect Your Herd.* Storey Publishing, 2007.

Ekarius, Carol. *How to Build Animal Housing.* Storey Publishing, 2004.

———. *Storey's Illustrated Breed Guide to Sheep, Goats, Cattle, and Pigs.* Storey Publishing, 2008.

Elliot, J., D. E. Lord, and J. M. Williams, eds. *British Sheep & Wool.* British Wool Marketing Board, 1990.

Hall, Stephen J. G., and Juliet Clutton-Brock. *Two Hundred Years of British Farm Livestock.* British Museum, 1989.

Henson, Elizabeth. *British Sheep Breeds.* Shire Publications, 1986.

+ Kleinheinz, Frank. *Sheep Management: A Handbook for the Shepherd and Student.* Madison, WI: Frank Kleinheinz, 1911.

+ Martin, W. C. L. *The Sheep: Our Domestic Breeds, and Their Treatment.* London: George Routledge & Co., 1852.

Müller, Hans Alfred. *Sheep: A Complete Owner's Manual.* Barron's, 1989.

Knight, Anthony P., and Richard G. Walter. *A Guide to Plant Poisoning of Animals in North America.* Teton NewMedia, 2001.

Parker, Ron. *The Sheep Book: A Handbook for the Modern Shepherd,* rev. ed. Swallow Press, 2001.

Pryor, Karen. *Don't Shoot the Dog! The New Art of Teaching and Training,* rev. ed. Ringpress Books, 2002.

Pugh, D. G., and A. N. Baird. *Sheep and Goat Medicine,* 2nd ed. Saunders, 2012.

+ Scott, John, and Charles Scott. *Blackface Sheep: Their History, Distribution, and Improvement.* Edinburgh: Ballantyne Press, 1888.

Scott, Philip R. *Sheep Medicine.* Manson, 2007.

Simmons, Paula, and Carol Ekarius. *Storey's Guide to Raising Sheep,* 4th ed. Storey Publishing, 2009.

Simmons, Paula. *Turning Wool into a Cottage Industry.* Storey Communications, 1991. First published in 1985 by Madrona Publishers.

+ Spooner, W. C. *The History, Structure, Economy and Diseases of the Sheep,* 3rd ed. London: Lockwood & Co., 1874.

+ Stewart, Henry. *The Domestic Sheep: Its Culture and General Management.* Chicago: American Sheep Breeder Press, 1898.

+ ———. *The Shepherd's Manual.* Orange Judd Company, 1876.

*Storey's Barn Guide to Sheep.* Storey Publishing, 2006.

Tillman, Peggy. *Clicking with Your Dog: Step-by-Step in Pictures.* Sunshine Books, 2000.

Weathers, Shirley A. *Field Guide to Plants Poisonous to Livestock: Western U.S.* Rosebud Press, 1998.

Weaver, Sue. *The Backyard Goat.* Storey Publishing, 2011.

———. *Sheep: Small-Scale Sheep Keeping.* BowTie Press, 2005.

———. *Storey's Guide to Raising Miniature Livestock.* Storey Publishing, 2010.

+ Wing, Joseph E. *Sheep Farming in America,* 3rd ed. Chicago: The Breeders Gazette, 1912.

+ Youatt, William. *Sheep: Breeds, Management, and Diseases.* London: Baldwin and Cradock, 1837.

WHO'S LITTLE

ARE · YOU ?

COPYRIGHT 1910 BY ROTH & LANGLEY. N.Y.

# Credits

## Illustrations Credits

© Elara Tanguy: 21, 27, 28, 29, 30, 31, 33, 40, 43, 44 top left, 46, 47, 68,70, 75, 76, 77, 83, 91, 95, 96, 97, 99, 105, 109, 124, 125, 133 top left, 142 top, 145, 146, 155, 161 left, 168, 169, 170, 171, 172, 178, and 192

© Elayne Sears: 13, 15, 16,17, 42, 44 all but top left, 45, 72, 73, 81, 93, 94, 106, 118, 130, 142 bottom, 147, 148–149, 151, 161 right, 162, 163, 179, 180, 181, 182, 183, 195

## Color Section Photography Credits

page 49: © Ronald Wittek/agefotostock

page 50: © Andrew Twort/Alamy, top left; © Niall McOnegal/Alamy, top right; © Wayne Hutchinson/ Alamy, middle left; © Ron Willbie/Animal-photography.com, middle right; © FLPA/Alamy, bottom left; © bickwinkel/Alamy, bottom right

page 51: © Alan Keith Beastall/Alamy, top left; © Jason Houston, top right; © Juniors Bildarchiv GmbH/ Alamy, middle left; © Factory Hill/Alamy, middle right; © Wayne Hutchinson/Alamy, bottom left; © Christine Whitehead/Alamy, bottom right

page 52: © Trinity Mirror/Mirrorpix/Alamy, top left; © FLPA/Alamy, top right; © Juniors Bildarchiv GmbH/Alamy, middle left; © Philip Perry/FLPA/Minden Pictures, middle center; © Wayne Hutchinson/Alamy, middle right; © tbkmedia.de/Alamy, bottom left; © imagebroker/Alamy, bottom right

page 53: © Christine Whitehead/Alamy, top left; © Danita Delimont Creative/Alamy, top center; © Eric Isselée/fotolia, top right; © Sue Weaver, bottom (all)

page 54: © Bailey Cooper Photography/Alamy, top left; © Superstock/Alamy, top right; © davidpage photography/Alamy, middle left; © EGimages/Alamy, middle right; © Sue Weaver, bottom left; © Geoff Simpson/Alamy, bottom right

page 55: © Juniors Bildarchiv GmbH/Alamy, top left, middle right; © Jason Houston, top right; © Sue Weaver, middle left, bottom left; © Mark Hitchen/Alamy, bottom right

page 56: © photomadnz/Alamy, top left; © Jorge Fernandez/Alamy, top right; © Bernd Mellman/Alamy, middle; © Robert Harding World Images/Alamy, bottom

page 57: © Icelandic Photo Agency/Alamy, top left; © Ozimages/Alamy, top right; © Tibor Bognar/ Alamy, middle left; © LOOK Die Bildagenturder Fotografen GmbH/ Alamy, middle right; © Mark Pearson/Alamy, bottom

page 58: © Mark Duffy/Alamy, top left; © Pegaz/Alamy, top right; © Richard Olsenius/Getty Images, bottom

page 59: © Lynne Silver/Alamy, top; © Paulette Sinclair/Alamy, middle left; © Arco Images GmbH/ Alamy, middle right; © Hemis/Alamy, bottom

page 60: © Viktor Kunz/123RF.com, top left; © Samum/123RF.com, top right; © Eric Isselée/fotolia.com, bottom left; Eric Isselée/iStockphoto.com, bottom right

page 61: © Eric Isselée/fotolia.com, top; © Exactostock/Superstock, bottom

pages 62–63: Mars Vilaubi

page 64: © Sue Weaver

# Index

Page numbers in *italic* indicate illustrations or photos; those in **bold** indicate tables.

## A

abomasum, 79, *81,* 81–82, 147

Abram (sheep), 119, 158

abscesses, **89,** *99,* 99–100

acidosis, 79, 98, 99

adrenaline (epinephrine), 90, **92,** 94, 96

Advance Lamb Milk Replacer, 145

*Adventures of Huckleberry Finn, The* (Twain), 86

Afghanistan, 4, *57*

Africa, 10

age of sheep, 17, *17,* 41

Ahrens, Paul, 159

Akbash dogs, *58*

alcoholic beverages (sheepy), **18**

alfalfa hay, 82, 106, 143

Algeria, *56*

Alice, Margot, 35

Allen, R. L., 8, 68

American Association of Small Ruminant Practitioners, 143

American Blackbellies, **24**

American Gotlands, **24,** 30

American Livestock Breeds Conservancy, 25, 28, 38

American Sheep Industry Association (ASIA), 143, 159, 164–65

American Veterinary Medical Association, 143

Americas and sheep, 8, 9, 11

ammonium chloride, 106–7

anaphylactic shock, **92,** 94, 96

ancient breeds, 12

Angel (sheep), *64,* 158

Angora Goat, *60*

animal guardians, *58–59,* 72–73, *72–73,* 75, 87, 110

antibiotics, 90, **92,** 100, 102, 103

Arcadia, 5

Argali sheep, 4, *4*

Argentina, **18**

Arizona, 166

Arthur (sheep), 137

Asia, 11–12

Asia Minor, 5

Asian Mouflons, 4, *4*

assisting at lambing, 131, 134–35

Australia, 10, **18,** 29, *57,* 121, 159, 161

Australian Sheep and Wool Show, 166

availability and buying sheep, 41

AVID, 46, 47

## B

Baabara (sheep), 140

Baalki (sheep), 152

*baas,* 15, 60

Baasha (sheep), 41, 119, 158

Babydoll Southdowns, **24,** 28, *28,* 32

Babylonian temple carvings, 12

*Backyard Goat, The* (Weaver), 177

*Bacteroides nodosus,* 100, 101

Badger-Faced Welsh Mountains, 30

Balwen Welsh Mountains, 30, *50*

banamine, **92**

bandaging supplies, **92**

banding tails, *142,* 142–43

*Banner Sheep Magazine,* 38

Barbados Blackbellies, **24,** 31, *31, 50, 61*

barbed wire fences, 71, 74

barber pole worms, 108

BBC News, 14, 98

beards, sheep vs. goats, 61, *61*

bedding, 69, 157

beer for peaked sheep, 83, *83*

being your own vet (sometimes), 90–97

belling a ram, 118

benign foot rot (hoof scald), 101

Bennett, Marty McGee, 75–76

Beulah Speckled Faces, 30

Bighorn Sheep, 4

Big Merino "Rambo" (largest statue), 121

BioClip, 161

birthing. *See* lambing

birthing dystocia, 41, 120, 136

Black-faced Cheviots, *2*

black-faced (colored-faced) breeds, 26, 27

Blackfaced Mountain Family, 25, 27, *27*

Black Hawaiians, 31

Blackies (Scottish Blackface), **24,** 25, *27,* 27, 33, 35, *51, 52, 64,* 98, 116, 155

blackleg, 100

Black Sheep Gathering (Oregon), 166

Black Welsh Mountains, **24,** 30, *30*

Blake, William, 115

Blakewells (Leicester Longwools), 28, 29

blind spot, 14–15, *15, 20*

bloat, 79, 90, 98, 99

blood stop powder, 34, **92,** 142, 192

blowflies, 111, 141

Bluefaced Leicesters, **24,** 25, 29, *50,* 169

bluetongue, 91

grain overload (acidosis), 79, 98, 99

Great Britain, 6–8, **7, 8,** 9, 12, **18,** 25, 27, 28, 29, 38, 85, 144, 166, 170

"Great Custom" (export tax on wool shipped to Europe), 6

Great Pyrennes, *59*

Greece, 167

Greyface Dartmoors, 29

guarantees, 40

guardians, animal, *58–59,* 72–73, *72–73,* 75, 87, 110

Guinness World Records, 121

Gulf Coast Natives, **24,** 29, *29*

gunshot, euthanasia, 195, *195*

Gwydion (sheep), 159

## H

*Haemonchus contortus,* 108

hair sheep, 31, *31,* 76, **88**

halters, 47, *47,* 75–76, *76,* **92**

Hampshire Downs, *50*

Hampshires, **24,** 26, 28, 84, 184

handspinners, 23, 26, 30, 154, 155, 161, 166

happy and healthy sheep, 48, 87

hauling sheep, 78

hay, 76–77, *77,* 79, **80,** 82, 83, 84, 87, 99, 106, 121, 191

head back delivery, 132, *132*

health, evaluating, 41

health records, 39

hearing of sheep, 15

heart rate, **89,** 91

heat cycles, 15, 16, 41, **116,** 120, 143

heat tolerance breeds, 23, **24**

heaviest newborn lamb, 121

Hebrideans, 30

"hefting" instinct, 38

Henry II (King of England), 6

Henry VIII (King of England), 6

herding sheep, 21

Herdwicks, 25, 38, *52*

heritage breeds, 23, 25

Herriot, James, 128

heterosis, 23

Hill Radnors, 30

hind-feet-first delivery, 130, *130*

history, sheep throughout, 3–12

Hobby Farms Sheep Yahoo Group, 24, 83, 191

Hoegger Supply Company, 69, 107

hog, hogg, hogget, hoggeral, **7**

Hog Islands, **24,** 29, 38

HomeAgain, 46, 47

*Home Cheese Making: Recipes for 75 Delicious Cheeses* (Carroll), 167

*Home Creamery, The* (Farrell-Kingsley), 167

Homer, 5

hoof rot, 38, 39, 87, 100–101, 117

hoof scald, 101

hooves

care of, 36, 37, 122

trimming, 183–84, *192,* 192–93

Hope (sheep), *64,* 137

Horned Dorsets, **24,** 28

horned sheep, 74, 77, 191

horns, 32–34, *33, 34,* 35, *52,* 61, *61*

Hottentots, 10

housing sheep, 3, 32, 36, 37, 48, 67–78, 157–58. *See also* feeding sheep

Howe, Jackie, 159

*"How Much Does Your Animal Weigh"* (University of Arizona Extension), 95

humans and sheep, 3

Hungary, *52,* 167

hypocalcemia (milk fever), 90, 101, 122

hypothermia, 90

## I

Ibo people of Nigeria, 11

Iceland, 11, *57,* 85, 144

Icelandics, 12, **24,** 26, 30, 33, *50, 53,* 116

Icelandic Sheep Breeders of North America, 43

Idaho, 166

identifying sheep, 42–47, *42–47*

Île de France, **24**

illness, signs of, 87, **88–89**

Imbolc festival, 144

IM (intramuscular) injections, 93, 94

India, *57*

indoor lambs, 150–51, *151,* 152

infectious keratoconjunctivitis (pinkeye), **88,** 102–3

injections, **92,** 93–95, 93–96, **94**

Inner Mongolia, 11

intelligence of sheep, 13–14

inter-digital dermatitis (hoof scald), 101

interior (cross) fences, 70

internal parasites, 68, 77, **88,** 107–10, *109*

intramuscular (IM) injections, 93, 94

introducing the sheep, 13–22

intuition, trusting your, 40

Iraq, 5, 72

Ireland, 27, 85

irritation signs, 21

Isabella (Queen of Spain), 9

Isle of Mann, *52*

Israel, **18**

Italy, 167

Ithaca, 5

ivermectin dewormers, 110

## J

Jacobs, 12, 25, 30, 31, 33, *51, 52,* 119

Jacobson's organ, 16

Japan, 12

Jarvis, William, 10

being your own vet
(sometimes), 90–97
buying your sheep, 36, 37
finding, 86–87, 191
lambing season, 135
Victoria, 166
Vikings, 4, 12
vinyl-coated fences, 70, *70*
vision of sheep, 14–15, *15*
vital signs, **89**
vitamin B1 (thiamine), **92,** 100
vitamin E, 121
vitamins, **80**
vomeronasal organ, 16

## W

Wales, 27
walker, 7
washing (scouring) fleece, 164–65
Washington, George, 9
water, **80,** 87, 106, 126
watering devices, 77–78
weaning dam-raised lambs, 143
Weaver, Sue, viii, 35, 118–19, 128,
137, 140, 152, 177
weaving, 6
wedder, 7

weighing sheep, *95,* 95
weight, **88**
welcker, **7**
Welsh Black Mountains, 12
Welsh Hill and Mountain Family,
25, *30,* 30
Welsh Hill Speckled Faces, 30
Welsh Mountains, *2,* 30, 98
Wensleydales, 25, 29, *51, 53,* 158
wethers, **7,** 41, 79, **80,** 87, 103,
105, 106, 161
White Cheese (recipe), 175
White Dorpers, 31
Whiteface Dartmoors, 29
white-faced breeds, 26, 27
white muscle disease, 107
white sheep, 5
*Whole Art and Trade of
Husbandry, The* (Googe), 3
whorls, 5
Wicklow Cheviot of Ireland, 27
wigging, **7**
wild sheep, 4, *4,* 4–5, *142*
William the I (King of England), 6
Wilmut, Ian, 98
Wiltshire Horns, 11, **24**
Wiltshire's Salisbury Plain, 6
wineries (sheepy), **18**

wire fences, 70, *70,* 71, 74
Wisconsin Sheep and Wool
Festival, 25
Wolcott, Mrs. Cliff, 66
Wolf Moon Wren (sheep), 128,
140
"wool churches," 6
Woolfest (Cumbria, England), 166
Wool Festival at Taos (New
Mexico), 166
wool fibers, 157
wool sheep, 5, 6, 9, 10, 11, 26, 68,
**88,** 154, 155, 191
Worshipful Company of
Woolmen, 6
wound treatments, **92**

## Y

Yan Tan Tethera, **8**
Year of the Sheep (Year of the
Goat), 12

## Z

Zawi Chemi Shanidar, Iraq, 5

# Other Storey Titles You Will Enjoy

BY THE SAME AUTHOR:

*The Backyard Cow.*
The essential guide to keeping a productive family cow, including basic instructions for beginners and experience-based insights for seasoned dairy farmers.
240 pages. Paper. ISBN 978-1-60342-997-9.

*The Backyard Goat.*
A fun, informative introductory guide to keeping productive pet goats.
224 pages. Paper. ISBN 978-1-60342-790-6.

*The Donkey Companion.*
A guide to selecting, breeding, training, enjoying, and caring for these friendly dependable animals.
352 pages. Paper. ISBN 978-1-60342-038-9.

---

*The Backyard Homestead Guide to Raising Farm Animals,*
    edited by Gail Damerow.
Expert advice on raising healthy, happy, productive farm animals.
360 pages. Paper. ISBN 978-1-60342-969-6.

*Chick Days*, by Jenna Woginrich. Photography by Mars Vilaubi.
A delightful photographic guide for absolute beginners chronicling the journey of three chickens from newly hatched to fully grown.
128 pages. Paper. ISBN 978-1-60342-584-1.

*Homegrown Honey Bees*, by Alethea Morrison. Photography
    by Mars Vilaubi.
A beginner's guide to your first year with bees, from hiving to harvest.
160 pages. Paper. ISBN 978-1-60342-994-8.

These and other books from Storey Publishing are available wherever quality books are sold or by calling 1-800-441-5700.
Visit us at *www.storey.com* or sign up for our newsletter at *www.storey.com/signup*.